T0262011

Aquaculture and the Environment

Edited by **Roger Creed**

New York

Published by Callisto Reference,
106 Park Avenue, Suite 200,
New York, NY 10016, USA
www.callistoreference.com

Aquaculture and the Environment
Edited by Roger Creed

© 2015 Callisto Reference

International Standard Book Number: 978-1-63239-080-6 (Hardback)

This book contains information obtained from authentic and highly regarded sources. Copyright for all individual chapters remain with the respective authors as indicated. A wide variety of references are listed. Permission and sources are indicated; for detailed attributions, please refer to the permissions page. Reasonable efforts have been made to publish reliable data and information, but the authors, editors and publisher cannot assume any responsibility for the validity of all materials or the consequences of their use.

The publisher's policy is to use permanent paper from mills that operate a sustainable forestry policy. Furthermore, the publisher ensures that the text paper and cover boards used have met acceptable environmental accreditation standards.

Trademark Notice: Registered trademark of products or corporate names are used only for explanation and identification without intent to infringe.

Printed in the United States of America.

Contents

Preface

The main aim of this book is to educate learners and enhance their research focus by presenting diverse topics covering this vast field. This is an advanced book which compiles significant studies by distinguished experts in the area of analysis. This book addresses successive solutions to the challenges arising in the area of application, along with it; the book provides scope for future developments.

Aquaculture is the science of cultivating aquatic animals and plants in fresh or marine waters. It is the extended version of fishing, derived from the fact that harvests of wild sources of fish and other aquatic species cannot keep up with the demand of an increasing human population. Amplification of aquaculture can result in less concern for the environment and therefore, the first prerequisite for sustainable aquaculture is clean water. However, poor management of aquatic species production can alter or even destroy existing wild habitats, increase local pollution levels or negatively impact local species. Keeping this in mind, aquatic managers along with scientists are searching for modern and more efficient solutions to various issues regarding fish farming. This book offers recent research outcomes on the relationship between aquaculture and environment, and consists of case studies from all over the world with the goal of improving and performing sustainable aquaculture.

It was a great honour to edit this book, though there were challenges, as it involved a lot of communication and networking between me and the editorial team. However, the end result was this all-inclusive book covering diverse themes in the field.

Finally, it is important to acknowledge the efforts of the contributors for their excellent chapters, through which a wide variety of issues have been addressed. I would also like to thank my colleagues for their valuable feedback during the making of this book.

Editor

Part 1

Sensitive Ecosystems:
Mangroves, Lagoons, Reefs

1

Impact of Shrimp Farming on Mangrove Forest and Other Coastal Wetlands: The Case of Mexico

César Alejandro Berlanga-Robles[1], Arturo Ruiz-Luna[1]
and Rafael Hernández-Guzmán[2]
*[1]Centro de Investigación en Alimentación y Desarrollo A. C.,
Unidad Regional Mazatlán
[2]Posgrado en Ciencias del Mar y Limnología, UNAM
Mexico*

1. Introduction

Since the middle of the twentieth century, the shrimp farming industry has shown steady growth along the tropical and subtropical coasts of the world. The world's cultivated shrimp production in 1950 was 1325 tons, amounting just 0.3% of the total production for these crustaceans, which were mainly extracted from coastal and estuarine environments. Thirty years later, by 1982, the global shrimp production surpassed one million tons. By 2009, shrimp production grew to nearly 3.5 million tons valued at approximately 14.6 billion dollars, amounting to 34% of the world's shrimp production, including marine and estuarine catches (Fig. 1) (FAO, 2011).

This escalation has seen intense debate regarding the economic, social and, particularly, environmental impacts produced by this activity. There is special concern for wetland losses, increased organic loading in coastal waters, the introduction of exotic species and the dispersal of harmful diseases (Boyd and Clay, 1998; Primavera, 2006).

The most controversial impact of shrimp farming is related to habitat loss. One of the main concerns is the deforestation of mangrove, a coastal vegetation type recognized as a highly productive shelter habitat for many commercial aquatic species. It has been estimated that between 1.0 and 1.5 million hectares of the world's coasts are covered by some type of shrimp farming (extensive, semi-intensive or intensive systems), and between 20 and 40% of this area is blamed as a cause of mangrove loss (Primavera, 2006). Thailand is considered to be an extreme example of this problem, as mangrove cover in this country was halved from 1960 to 1996. Approximately 200,000 ha of mangroves were deforested, with a third of the area being transformed into shrimp farming ponds (Aksornkoae & Tokrisna, 2004).

Although shrimp farming impacts have been widely documented and discussed, there is little evidence on the real mangrove deforestation rates at regional or national scales due to this activity. Thus, some of the global estimates on mangrove deforestation for shrimp pond construction are imprecise projections based on very local studies or generalizations of extreme cases such as Thailand.

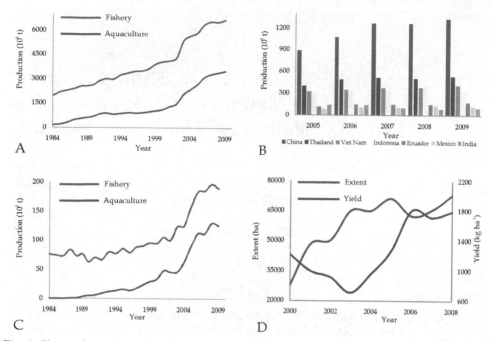

Fig. 1. Shrimp farming indicators: A) World shrimp production (1984-2009). B) Main producing countries (2005-2009). C) Mexican shrimp farming production (1984-2009). D) Extent and annual yield in Mexico (2000-2008).

The objective of the present study was analyze the land use changes caused by shrimp farming in the coastal landscape of Mexico, one of the main producers worldwide, using remote sensing (RS) and geographic information system (GIS) tools within a landscape change framework to contribute to a better understanding of the impacts of shrimp farming on coastal wetlands. The results were then compared with others obtained at different latitudes to gain a more precise knowledge of the responsibility of shrimp farming on mangrove deforestation and other environmental impacts.

1.1 Shrimp farming in Mexico

Shrimp farming has its origin in the late nineteenth century, but it was not until the 1960s and early 1970s that it became a commercial activity (Kungvankij et al., 1986). Mexico followed a similar trend, starting shrimp production in the early 1970s with the operation of an experimental farm to the northwest. However, legal issues related to land tenure complicated this development, particularly for private investments, until the middle 1980s, when laws changed, allowing the expansion of commercial farms, mainly in the same region.

Thereafter, like the rest of the world, shrimp farming in Mexico displayed rapid evolution, growing from 35 t of shrimp production in 1985 to 125,778 t in 2009 (Fig. 1). Profits also increased, from $ 175,000 to 405 million dollars, respectively. The net income by farms in northwest Mexico (semi-intensive systems) has been estimated between US$1.2 and US$2.9

per kg, with feed and seed prices as the major constraints for investors (Ponce-Palafox et al., 2011). However, farmed shrimp production has grown from 0.05% to 40% of the total national production for this crustacean (FAO, 2011), and Mexico is currently positioned among the ten largest producers of farmed shrimp in the world (Fig. 1) (Conapesca, 2009; 2010).

At the country level, shrimp aquaculture is practiced in almost all 17 coastal states. Even in inland locations, there are some initiatives to cultivate the same species used in marine aquaculture but adapted to freshwater environments. Although shrimp aquaculture is widespread nationally, the Gulf of California region is the most highly concentrated region of activity, with the states of Sonora, Sinaloa and Nayarit representing more than 95% of the total shrimp pond extent and production in Mexico. By contrast, Jalisco, Michoacán, Oaxaca, Chiapas and Tabasco together amount to less than 1% (Fig. 2) because physiographic or economic factors have inhibited the development of this activity.

Some species of the genera *Litopenaeus* and *Farfantepenaeus* have been used for commercial purposes, but the white shrimp *L. vannamei* is currently the most common species in culture. This species is grown in one (8-9 months) or two cycles (3-4 months each) a year, obtaining a final weight between 10 to 25 g in the first case and 7 to 11 g in the second. Even when the use of wild postlarvae (PL) is allowed in Mexico, with permission granted for extraction, this activity is sustained by PL production controlled in 33 laboratories that produce in average of approximately 76 million PL per year. The last reliable record of aquaculture in Mexico (CONAPESCA, 2010) states a total output of approximately 72 900 ha as of 2008 (Figure 1). In almost all cases, the shrimp farms use semi-intensive production systems, which, aside from the certified larvae, require substantial amounts of fertilizers to increase natural productivity and complementary feed to maintain stocking densities from 6 to 30 postlarvae per area (PL/m^2).

With this system, and considering the figures on total shrimp pond area and production, the average yield from 2000 to 2008 was 1260 kg ha^{-1} (Fig. 1), although it was lower from 2000 to 2003, when sanitary problems associated with viral diseases occurred, increasing later to approximately 1750 kg ha^{-1}, a level that has been maintained since 2006 (CONAPESCA, 2009; 2010). In agreement with Ponce-Palafox et al. (2011), the top three producer states in Mexico obtained average yields of 800 (Nayarit), 900 (Sinaloa) and 3200 (Sonora) kg ha^{-1} per crop.

2. Methods: Land use changes associated with shrimp farming in Mexico

To analyze the land use changes caused by shrimp farming in Mexico and to estimate rates of coastal wetland loss induced by this activity, we performed a change detection analysis in three steps following a procedure similar to that proposed by Berlanga-Robles et al. (2011). Because shrimp farming in Mexico is concentrated around the northern states, particularly the east coast of the Gulf of California, to make this study representative, four states that account for 97% of this activity in extent and production were chosen for the analysis: Nayarit, Sinaloa and Sonora in the Gulf of California and Tamaulipas in the Gulf of Mexico (Figure 2).

2.1 Shrimp farm location and inventory

First, the shrimp farms of the four states selected were geographically located with a database provided by the National Commission for Fisheries and Aquaculture (CONAPESCA).

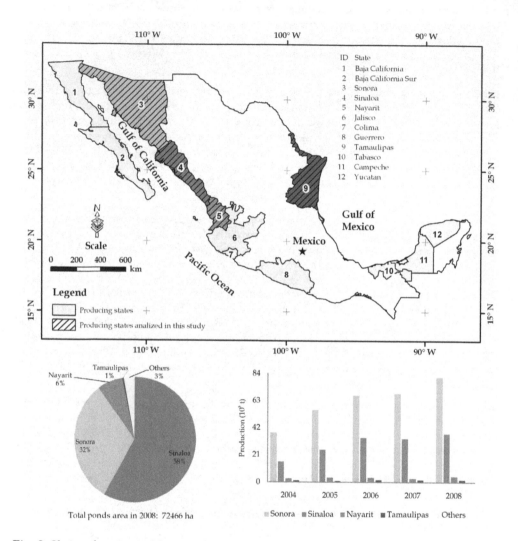

Fig. 2. Shrimp farming in Mexico. The bar graph shows the proportion of pond area by state and the bar graph shows production by state from 2004 to 2008.

This database was updated and corrected by visual interpretation of the Quickbird and GeoEye imagery available on Google Earth (2002 to 2011) as well as false-color composites from Landsat TM (2010 and 2011) and SPOT panchromatic (2010) imagery with a 30 and 2.5 m pixel size (Figure 3).

When the polygons in the four states were completed, a 500 m buffer zone was created around them using geographic information system (GIS) tools. The farms' area and their buffer zones were then used to mask the Landsat TM images used in the next step so that the area outside of them formed a background without spectral information.

Step 1: Shrimp farms inventory

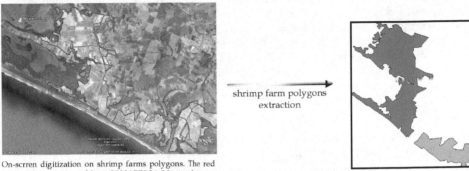

shrimp farm polygons
extraction

On-scrren digitization on shrimp farms polygons. The red
polygons were imported from CONAPESCA-Mexico dataset.
In blue, polygons corrected and updated from GeoEye and
QuickBird imagery (2010-2011) in Googe Earth.

Step 2: Classification of Landsat TM images (1986-1990)

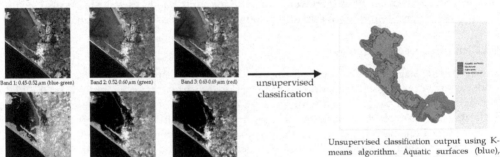

Band 1: 0.45-0.52 μm (blue-green)   Band 2: 0.52-0.60 μm (green)   Band 3: 0.63-0.69 μm (red)

Band 4: 0.76-0.90 μm (near infrared)   Band 5: 1.55-1.75 μm (mid infrared)   Band 7: 2.08-2.35 μm (mid infrared)

unsupervised
classification

Unsupervised classification output using K-
means algorithm. Aquatic surfaces (blue),
mangrove (green), saltmarsh (grey) and
terrestrial covers (brown).

Spectral bands of sensor Thematic Mapper (TM) carried on
satellites Landsat 4 and 5 used in this study. Coast of Sinaloa
(path/row: 32/43).

Step 3: Land use change detection

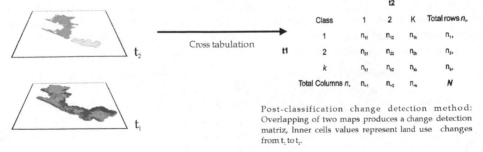

Cross tabulation

		t2			
	Class	1	2	K	Total rows n_{i*}
t1	1	n_{11}	n_{12}	n_{ik}	n_{1*}
	2	n_{21}	n_{22}	n_{2k}	n_{2*}
	k	n_{k1}	n_{k2}	n_{kk}	n_{k*}
	Total Columns n_{*j}	n_{*1}	n_{*2}	n_{*k}	N

Post-classification change detection method:
Overlapping of two maps produces a change detection
matriz, Inner cells values represent land use changes
from t_1 to t_2.

Fig. 3. Technical process to detect and assess landscape changes produced by shrimp ponds
construction based on satellite imagery analysis and ancillary data.

2.2 Landscape characterization

In the second step, performed by analysts different than those whose updated and prepared the shrimp farm polygons, the coastal landscape of four selected states before the advent of shrimp farming were characterized by means of thematic maps generated by the classification of Landsat TM images from 1986 to 1999, downloaded from the USGS Global Visualization Viewer portal (http://glovis.usgs.gov/). The imagery covering the shrimp farming area in the states of the Gulf of California comprises 14 Landsat TM images among paths 30 to 37 and rows 39 to 45. The area of Tamaulipas was covered with three images recorded in path 26 among rows 41 and 43. All the spectral bands except thermal infrared were used.

The images underwent unsupervised classification using a K-means clustering technique (Richards & Jia, 1999). A 16 spectral cluster map was produced first, which was subsequently associated with natural covers represented by three coastal wetland types (aquatic surfaces, saltmarsh and mangroves), while other natural vegetation (dry forest, thorn scrub forest) and vegetation of anthropic origin (agriculture, settlements, lineal infrastructure) were integrated into a fourth category: terrestrial covers (Fig. 3). Landsat TM images recorded earlier than 1986 were not available, so in some cases the maps also include a fifth land cover category corresponding to the shrimp farms present since that time.

2.3 Change detection analysis

In the third step, the changes produced by shrimp farming in the Mexican coastal landscapes were assessed by overlaying the buffered shrimp farm polygons (t2) on the 1986-1999 thematic maps (t1) following a post-classificatory analysis scheme (Mas, 1999, Berlanga-Robles et al. 2010; Berlanga-Robles & Ruiz-Luna, 2011), which outputs a matrix for change detection, identifying trends and the extent of variations on every cover presumably produced by shrimp farming (Fig. 3). Considering just the Gulf of California region, a similar analysis was performed only on mangroves, using a dataset produced with 1973 Landsat MSS images (60 m pixel size) developed in earlier studies (Ruiz-Luna et al., 2010).

3. Results

Based on the photo-interpretation process with Google Earth and ancillary data, a total of 273 polygons were identified, representing isolated farms or systems with more than one farm amounting to a total of approximately 80,000 ha. All structures identified as shrimp farms were included, even if the system was empty or out of operation. Sinaloa state has the largest area allocated for this industry, amounting to 51% of the estimated area, followed by Sonora, Nayarit and Tamaulipas, with 41, 6 and 1%, respectively (Table 1).

Regarding the transformations due to aquaculture, the main subsidiaries were those that integrate anthropic and vegetation cover other than that identifiable as wetland, namely, terrestrial coverage, with 46%, and saltmarsh, with 45%. Approximately 3% of the ponds were built on the shallow coastal lagoons and estuaries (water surface), and mangrove was the least modified cover (1%). The change in mangrove cover is estimated to be more than 1150 ha, mainly in Sinaloa ($\approx$ 700 ha) and Nayarit ($\approx$ 400), the states with the largest mangrove cover in the Mexican Pacific, which account for approximately 70,000 ha each (Ruiz-Luna et al., 2010). These states are also first in the execution of shrimp aquaculture

| Subsidiary cover | State | | | | |
	Nayarit	Sinaloa	Sonora	Tamaulipas	TOTAL
Aquatic surfaces	103 (2)	1918 (5)	268 (1)	24 (1)	2313 (3)
Mangrove	392 (8)	689 (2)	85 (<1)	0	1166 (1)
Saltmarsh	1726 (35)	23225 (56)	10779 (33)	48 (6)	35778 (45)
Terrestrial covers	2507 (50)	12215 (30)	21133 (64)	929 (93)	36784 (46)
Shrimp farming*	238 (2)	3090 (2)	642 (2)	0	3970 (5)
Total shrimp farm extent (ha)	4966	41137	32907	1001	80011

Table 1. Land use changes produced by shrimp farming in four states of Mexico. Area in hectares and corresponding proportion (%) in parenthesis. *Some farms were built before 1986, the initial time for this study (t1), consequently, figures in this row represent no change after this date.

projects. It is important to highlight that shrimp aquaculture started prior to 1986, the date of the first Landsat image included in this analysis, which explains why 5% of the shrimp aquaculture use was unchaged in land use . It means that approximately 4000 ha of shrimp ponds had been constructed by 1986 on undetermined covers.

As most of the changes happened in the Gulf of California region, it is important to have look at Tamaulipas, the only representative of shrimp aquaculture in the Gulf of Mexico. No mangrove deforestation was associated with shrimp ponds, and the main subsidiary cover was terrestrial cover, amounting to 93% of the total area used for shrimp pond installation.

Based on the 1973 estimates for mangrove distribution proposed by Ruiz-Luna et al. (2010) for the Gulf of California region, the change detection analysis output some differences with the previous analysis, showing a slight reduction of the assessed mangrove loss for Nayarit and Sonora (Table 2). The mangrove change at Nayarit was 77 ha less, as evaluated with the 1973 map with respect to the 1986 map. The changes in Sonora were similar in both studies; even so, the reduction is 14 ha more with the 1986 map than that estimated with the 1973 map. The differences in both cases are approximately 15-20%. By contrast, the mangrove loss estimated for Sinaloa increased by approximately 40% when the 1973 map was analyzed, agreeing with a technical report published by Ruiz-Luna et al. (2005). Even so, the technical differences in both Landsat devices (MSS and TM) make an underestimation of the 1973 mangrove area possible due the low resolution of the Landsat MSS imagery (60 m) used to produce these maps, as noted by Ruiz-Luna et al. (2010). Thus, the differences among Nayarit and Sonora could be reduced, but in the case of Sinaloa, it could increase.

| | State | Land covers (1973) | | Total |
		Mangrove	Others covers	
Shrimp farms (2010)	Nayarit	315 (6)	4651 (94)	4966
	Sinaloa	956 (2)	40181 (98)	41376
	Sonora	71 (<1)	32836 (>99)	32907

Table 2. Change detection matrix for land cover change from mangrove (1973) to shrimp farms (2010). Area in hectares and relative proportion (%) in parenthesis.

Mexican laws protect mangrove forests (Federal Wildlife Law 2000, NOM-022-SEMARNAT-2003, NOM-059-SEMARNAT-2010). These laws declare all mangrove species endangered, and they forbid changes on this cover while prohibiting adjacent economic activities (with some exemptions). Therefore, we defined a 100-m buffer zone around the shrimp farm polygons to assess the impact on mangroves within this fringe restricted by law. Using this criterion, the impacted area is almost twice the preceding figure, and Sinaloa was again the most unsafe area. The results of this analysis are shown in Table 3.

State	Shrimp farm polygons	% of farms adjacent	Mangrove in 100 m zone
Nayarit	43	81.4	426
Sinaloa	163	68.7	1635
Sonora	45	26.7	113
Tamaulipas	22	4.5	2
Total	273	58.6	2176

Table 3. Shrimp farms adjacent to mangrove forests in some Mexican states and mangrove extent (ha) inside the 100 m fringe banned by law for any economic activity. Estimations are based on a 100 m buffer created around the shrimp farm polygons and overlaid on mangrove thematic maps.

Even when a farm's design excluded the polygon from the mangrove cover, the shrimp farms were sometimes constructed in the vicinity of mangroves, thus transgressing some environmental regulations. From our results, close to 60% of the total analyzed shrimp farms were in proximity with mangroves, almost doubling the lost area estimated here for this vegetation if the 100 m fringe is considered. Sinaloa and Nayarit, both with the largest mangrove coverage, were the states with the highest interaction between polygons and the forbidden perimeter, affecting more than 80% of the farms in the case of Nayarit.

4. Mexican shrimp farming in the international context

Comparing the observed conditions of Mexican shrimp farming with other producing countries worldwide highlights the fact that most of the declarations about mangrove deforestation by shrimp farming are not properly documented. Documents with data and descriptions of the technical process to assess mangrove deforestation are limited and, in some cases, only generalize observed trends. From the available information, it was possible to analyze some cases from Asia and America, including six cases in Mexico (Table 4).

At first sight, the situation of the Mexican states can be roughly compared with that from other countries. In some regions of India, Bangladesh and Vietnam, though not necessarily at the country level, shrimp ponds are practically the only cause of deforestation, with rates between 85 and 100%, even when those ponds generally represent a small to medium fraction (17.6 to 53%) of the activity (Table 4).

Shrimp farming growth in Latin America also had negative effects on mangrove cover, particularly in Ecuador and Honduras (Gulf of Fonseca), with a decline in total mangrove cover of approximately 27% and 22% between 1969-1995 and 1973-1992, respectively (DeWalt et al., 1996; Tobey et al., 1998). These references agree with the present analysis,

Country/Region (period)	Mangrove area (ha)		A	B	C		
	t_1	t_2	Loss	Ponds	Conv	%1	%2
Chakaria Sunderban, Bangladesh (1952-1988)[1]	7500	978	6522	6522	6522	100	100
Thailand (1961-1996)[2]	367900	167582	200318	80000	66998	84	33
Tien Hai, Viet Nam (1986 1992)[3]	1250	390	860	1490	790	53	92
Philippines (1997)[a,4]	295000[17]	250000[17]	45000	6940	3470	50	8
Ecuador (1969-1999)[5]	362700	149557	213143	10000	48649	49	23
Machala-Pto. Bolivar, Ecuador (1966-1982)[6]	4693	3294	1399	2331	931	40	67
Giao Thuy, Viet Nam (1986-1992)[3]	750	320	430	938	364	39	85
Golfo de Fonseca, Honduras (1973-1992)[7]	30697	23937	6760	11515	4307	37	64
Indonesia (1997)[a,4]	4200000[17]	3150000[17]	1050000	20000	5320	27	1
Thailand (1997)[a,4]	280000[17]	244100[17]	35900	47755	9933	21	28
Godovari delta, India (1977-2005)[8]	19480	18610	870	4650	820	18	94
Taiwan (1997)[a,4]	--	--	--	1407	173	12.3	--
San Blas, Mexico (1973-2001)[9]	7644	7353	521	1900	230	12	44
CaiNuoc district, Viet Nam (1968-2003)[10]	19507	47614	14746	59684	5643	10	38
Mazatlan, Mexico (1973-1997)[11]	910	710	200	170	10	6	5
India (1989-1999)[12]	467000[17]	448200[17]	18800	130000	6500	5	35
Marismas Nacionales, Mexico (1973-2000)[13]	89182	75042	14140	3208	102	3	1
Sinaloa, Mexico (1992-2003)[14]	75364	84912	--b	46882	790	2	--
Ceuta Lagoon, Mexico (1984-1999)[15]	7558	7217	341	3192	23	1	7
North of Sinaloa (1986-2005)[16]	21983	21873	110	9949	26	<1	24

t1 and t2, initial and final mangrove area in every period study. A Loss, differences in mangrove cover (ha) in the study period. B Ponds, area (ha) occupied by shrimp ponds. C Conv., conversion from mangrove to shrimp ponds (area in ha). %1, proportion of ponds built on mangrove = (C/B)100. %2, proportion of mangrove loss caused by shrimp farming = (C/A)100. [a] Only considering intensive shrimp farms. [b] Authors found a positive change in Sinaloa's mangrove cover. Sources: [1]Hossain (2001); [2]Aksornkoae and Tokrisna (2004); [3]Bélard et al. (2006); [4]Kongkeo (1997); [5]Bravo (2003); [6]Terchunian et al. (1986); [7]Dewalt et al. (1996); [8] Sudhaka-Reddy & Roy (2008); [9]Berlanga-Robles & Ruiz-Luna (2006); [10]Binh et al. (2005); [11]Ruiz-Luna & Berlanga-Robles (2003); [12]Hein (2002); [13]Berlanga-Robles & Ruiz-Luna (2007); [14]De la Fuente & Carrera (2005); [15]Alonso-Pérez (2003); [16]Berlanga-Robles et al. (2005); [17]FAO (2007).

Table 4. Changes in mangrove cover related to shrimp farming in Asia and America.

which reveals that Ecuador's mangrove loss and conversion was close to 90% of approximately 54000 ha at the nation level and 67% at the regional level (Machala-Puerto Bolivar), where mangrove loss was estimated at approximately 1400 ha. Honduras in the early 1970s accounted for more than 11500 ha of shrimp ponds in the Gulf of Fonseca, approximately 65% of which were constructed on mangrove sites.

In Mexico, the highest conversion ratio from mangrove to shrimp ponds has been recorded for San Blas, at Nayarit state, and northern Sinaloa, with 44.1% and 23.6%, respectively, amounting to a total of approximately 260 converted hectares. In relative terms, these numbers represent 12.1% and 0.3% of the total pond area constructed by region, respectively, at approximately 12 000 ha in total. Other studies in Mexico on mangrove conversion attributable to shrimp farming show output ratios less than 7% of lost mangrove cover. It is also remarkable that two independent works conducted in Sinaloa state found an increase in the mangrove cover, and, with some differences, they even found that the mangrove area occupied by shrimp farm developments represents between 1.7 and 2.2% of a pond surface estimated above 40 000 ha (De la Fuente & Carrera, 2005; Ruiz-Luna et al., 2005).

5. Discussion

Intense debate about the environmental impacts caused by shrimp farming has been engaged in Mexico since the beginning of this activity, particularly by environmentalists, regarding the denunciation of the environmental risks associated with shrimp farming development. Considering the international background of this issue and bearing in mind the importance of the environmental services offered by mangroves and the possible impact caused by land cover changes, the general opinion is that Mexico could confront environmental risks similar to Indonesia, Philippines, Thailand and Ecuador, where extensive deforestation of mangrove forests is associated with the construction of shrimp ponds.

This perception has been maintained and consistently declared even though there are few studies documenting changes in mangrove at a national extent or the possible causes of mangrove deforestation when it is proved. It is common that some differences in mangrove cover estimations obtained by the extrapolation of local values or using different inputs and evaluation techniques would be misinterpreted as deforestation (Ruiz-Luna et al. 2008).

Thus, the studies conducted by Hernández-Cornejo & Ruiz-Luna (2000), Alonso-Pérez et al. (2003), De la Fuente & Carrera (2005), Ruiz-Luna et al. (2005), Berlanga-Robles & Ruiz-Luna (2007), and Berlanga-Robles et al. (2011), among others, have attempted to verify the extent and intensity of the impact of shrimp farming in Mexico.

Most of the above papers mainly describe the conditions observed in Sinaloa and Nayarit, in northwest Mexico. This paper is the first attempt to document changes at a nationwide level based on our own and other authors findings obtained with remote sensing techniques, analyzing very high spatial resolution satellite imagery (Landsat, Spot, QickBird, GeoEye) and updating the existing information up to 2010. The main restriction imposed to these studies is the lack of reference data to validate the accuracy of the earlier dates' estimates. Even so, the similitude among the results from independent analyses give confidence to the general trends followed by shrimp farm growth and its impact on mangrove forests in Mexico, making a comparison possible with analogous developments elsewhere.

We must emphasize that, even with the largest mangrove extent and the best developed area being located in the Yucatan Peninsula (Campeche, Yucatan, Quintana Roo and Chiapas states), which accounts for approximately 60% of the mangrove forests in Mexico (Acosta et al., 2009), the shrimp farming in this area represents less than 1% of the total extent and production of Mexico (CONAPESCA, 2010). For this reason, neither of the abovementioned states were included in the analysis. The four states analyzed here currently amount to 97% of the area dedicated to shrimp farming (CONAPESCA, 2010), which is enough to document the impact of this activity on mangrove cover.

The present findings indicate that Mexico has approximately 82 500 ha dedicated to shrimp production, though not all of this area is necessarily in operation. From these areas, between 1.5% and 1.7% could be constructed on mangrove cover, removing approximately 1300 ha, which is equivalent to less than 1.0% of the 770 000 ha of mangrove reported by the Mexican National Commission for the Knowledge and Use of Biodiversity (Acosta et al., 2009). These results greatly contrast with other tropical and subtropical countries, where shrimp farming has been responsible for of most of the mangrove deforestation (Bangladesh, Ecuador) or an important part of it (Honduras, India, Thailand). However, although shrimp farming could not be considered a risk to Mexican mangrove cover, it has been established on other important coastal wetlands rarely mentioned in literature (estuaries, lagoon, saltmarsh).

The worldwide estimation of mangrove deforestation caused by shrimp farming is difficult because not all producing countries have reliable data at the national level. The analysis of the literature shows that in many instances, nationwide or global estimates are based on local or regional case studies or are extrapolated from foreign conditions, such as those from Thailand and Ecuador, or even Indonesia, where mangrove loss has been severe though mostly independent of shrimp farming activity.

In agreement with FAO (2007), the global mangrove cover declined from approximately 19 million ha in 1980 to almost 16 million ha in 2000, while the shrimp pond area was 1.25 million ha in 1998 (Rönnbäck, 2002). Considering the extreme case of all the shrimp ponds constructed on mangroves areas, this activity could be responsible for 41% of mangrove loss. As observed here, in approximately 70% of the cases, the shrimp farming accounted for less than 50% of deforestation, and within this 70%, the half has contributed with less than 30% of mangrove decline. Considering both scenarios, shrimp farming could be directly responsible for 20.8 to 12.5% of the mangrove loss between 1980 and 2000.

The Mexican case could be a result of a postponed development of the industry, with a delay of approximately 10 to 15 years in respect to other countries due to legal constraints. After this late beginning, the industry grew rapidly even while acknowledging environmental problems and is now among the ten top producers, second to Latin America, which is after Ecuador. Consequently, shrimp farming has been responsible for mangrove deforestation but not at the same level observed in the former shrimp producers. Regrettably, the risk has been transferred to other coastal wetlands, as 46% of the ponds have been built on saltmarshes. This land cover is more suitable for shrimp pond construction farms because of soil characteristics and topography. In addition, these wetlands are cheap in economic terms, as they are considered unproductive, and they are barely protected by Mexican laws. Studies on saltmarsh loss show that 12% of this cover in Nayarit and Sinaloa was lost because of 25 000 ha of ponds (Berlanga-Robles et al., 2011). Even more, the impact of shrimp farming on the coastal landscapes goes beyond the direct loss of wetlands because the ponds themselves and mostly the linear infrastructure necessary for the operation of farms, such as canals and roads, have a strong impact on the connectivity of coastal landscapes, fragmenting saltmarsh habitat, modifying the water flows and sediment supplies in the intertidal zone, and threatening the overall stability of coastal wetlands (Berlanga-Robles et al., 2011).

In conclusion, the shrimp aquaculture in Mexico is not the main cause of mangrove deforestation, as has happened with other countries. Even so, the industry is far from sustainable because almost half of the pond area has resulted in the direct removal of other natural wetlands. Also, the entire associated infrastructure interrupts local and regional ecological process by fragmentation of the intertidal zone (Berlanga-Robles et al., 2011). Finally, even when those farms do not have contact with the mangrove cover, a significant proportion of them were built near mangrove patches, particularly in Sinaloa and Nayarit, infringing upon legal rules and threatening the 100 m fringe established by Mexican law. To move toward real sustainability, some areas must be restored in agreement with laws. Future developments must require an ecologic and economic reevaluation of coastal wetlands prior to operation to avoid new impacts and to provide the systems with the essential connectivity among wetlands and other wetlands, maintaining the water and sediment flows in the intertidal zone.

6. Acknowledgment

The authors acknowledge the National Commission for Fisheries and Aquaculture-Mexico (CONAPESCA) for provided us his database with the shrimp farms polygons. Landsat images used in this study were provided by U. S. Geological Survey Resources Observation and science Center (USGS-EROS).

7. References

Acosta-Velázquez, J., Rodríguez-Zuñiga, T., Reyes-Díaz-Gallegos, J., Cerdeira-Estrada, S., Troche-Souza, C., Cruz, I., Ressl, R. & Jiménez, R. (2009) Assessing a nationwide spatial distribution of mangrove forest for Mexico: an analysis with high resolution images. *33rd International Symposium on Remote Sensing of Environment*, Stressa, Italy, 2009 may.

Aksornkoae S. & Tokrisna, R. (2004). Overview of shrimp farming and mangrove loss in Thailand, In: *Shrimp Farming and Mangrove Loss in Thailand*, Barbier, E.B. Sathirathai, S., pp. 37-51, Edward Elgar, ISBN 1843766019, Great Britain.

Alonso-Pérez, F., Ruiz-Luna, A., Turner, J., Berlanga-Robles, C.A. & Mitchelson-Jacob, M.G. (2003) Land cover changes in the Ceuta coastal lagoon system, Sinaloa, Mexico: assessing the effect of the establishment of shrimp aquaculture. *Ocean & Coastal Management*, 46, pp. 583-600.

Bélard, M., Goïta, K., Bonn, F. & Pham, T.T.H. (2006) Assessment of land-cover changes related to shrimp aquaculture using remote sensing data: a case study in the Gian Thuy Distric, Vietnam. *International Journal of Remote Sensing*, 27, pp. 1491-1510.

Hein, L. (2002). Toward improved environmental and social management of Indian shrimp farming. Environmental Management 29, 3, pp. 349-359.

Berlanga-Robles C.A. & Ruiz-Luna, A. (2006) Assessment of landscape changes and their effects on the San Blas estuarine system, Nayarit (Mexico), through Landsat imagery analysis. Ciencias Marinas, 32, 3, pp. 523-538.

Berlanga-Robles C.A. & Ruiz-Luna, A. (2007) Análisis de las tendencias de cambio del bosque de mangle del sistema lagunar Teacapán-Agua Brava, México. Una aproximación con el uso de imágenes de satélite Landsat. Universidad y ciencia, 23, 1, pp. 29-46.

Berlanga-Robles, García-Campos, R.R., López-Blanco J. & Ruiz-Luna, A. (2010) Patrones de cambio de coberturas y usos del suelo en la región costa norte de Nayarit (1973-2000). Investigaciones Geográficas, 72, pp. 7-22.

Berlanga-Robles, C.A., Ruiz-Luna A. Bocco G. & Vekerdy Z. (2011). Spatial analysis of the impact of shrimp culture on the coastal wetlands on the Northern coast of Sinaloa, Mexico. Ocean & Coastal Management, 54, pp. 535-543.

Binh T.N.K.D., Vromant, N., Hung, N.T., Hens, L. & Boon E.K., 2005. Land cover changes between 1968 and 2003 in Cai Nuoc Ca Mau Peninsula, VietNam. Environment Development and Sustainability, 7, pp. 519-536.

Boyd, C.E. & Clay, J.W. (1998). Shrimp aquaculture and the environment. Scientific American, 278, 58-65.

Bravo, E. (2003) La industria camaronera en Ecuador, Globalización y Agricultura. Jornadas para la Soberanía Alimentaria, Barcelona, 2003 june.

CONAPESCA (2009) Anuario Estadístico de Acuacultura y Pesca Edición 2007. Comisión Nacional de Acuacultura y Pesca, Mexico.

CONAPESCA (2010) Anuario Estadístico de Acuacultura y Pesca Edición 2008. Comisión Nacional de Acuacultura y Pesca, Mexico.

De la Fuente L.G. & Carrera, G.E. (2005) Cambio de Usos del Suelo en la Zona Costera del Estado de Sinaloa, Ducks Unlimited de México, México.

DeWalt, B., P. Vergne, P. & Hardin, M. (1996). Shrimp aquaculture development and the environment: People, mangroves and fisheries in the Gulf of Fonseca. World Development, 24, 7, pp. 1193-12098.

FAO, (2007). The World's Mangroves 1980-2005, FAO Forestry Paper 153, ISBN 9789251058565, Rome.

FAO, (2011). Global Aquaculture Production 1950-2009, accessed 22-07-11, http://www.fao.org

Hernández-Cornejo R. & Ruiz-Luna, A. (2000) Development of shrimp farming in the coastal zone of southern Sinaloa (Mexico): operating characteristics, environmental issues, and perspectives. Ocean & Coastal Management, 43, 597-607.

Hossain, MdS., Lin, C.K, Hussain & M.Z. (2001) Goodbye Chakaria Sunderban: The oldest mangrove forest. The Society of Wetland Scientists Bulletin, 18, 3, pp. 19-22.

Kongkeo, H. (1997) Comparison of intensive shrimp farming systems in Indonesia, Philippines, Taiwan and Thailand. Aquaculture Research 28, pp 789-796.

Kungvankij, P., Chua, T.E., Pudadera, B.J. Jr., Corre, K.G., Borlongan, E., Alava, Tiro, L.B. Jr., Potestas, I.O. & Talean, G.A. (1986). Shrimp Culture: Pond Design, Operation and Management. FAO Training Series No. 2, http://www.fao.org

Mas, J.F. (1999) Monitoring land-cover changes: a comparison of change detection techniques. International Journal of Remote Sensing, 20(1), pp. 139-152.

Ponce-Palafox JT, Ruiz-Luna A, Castillo-Vargasmachuca S, García-Ulloa M & Arredondo-Figueroa J.L. (2011). Technical, economics and environmental analysis of semi-intensive shrimp (Litopenaeus vannamei) farming in Sonora, Sinaloa and Nayarit states, at the east coast of the Gulf of California, México. Ocean & Coastal Management, 54, pp. 507-513

Primavera, J.H. (2006). Overcoming the impacts of aquaculture on the coastal zone. Ocean & Coastal Management, 49, pp. 531-545.

Richards J.A., Jia, X. (1999) Remote Sensing Digital Image Analysis, Springer, ISBN 3540648607, Berlin.

Rönnbäck, P. (2002) Environmentally Sustainable Shrimp Aquaculture, Swedish Society for Nature Conservation, http://www.naturskyddsforeningen.se

Ruiz-Luna, A. & Berlanga-Robles, C.A. (2003) Land use, land cover changes and coastal lagoon surface reduction associated with urban growth in northwest Mexico. Landscape Ecology, 18, pp. 159-171.

Ruiz-Luna A., Acosta-Velázquez, J., Monzalvo-Santos, I.K. & Berlanga-Robles, C.A. (2005) *Evaluación de la cobertura de manglar, estructura forestal y determinación del impacto potencial por el establecimiento de granjas camaronícolas*, Instituto Sinaloense de Acuacultura (ISA-Sinaloa), México.

Ruiz-Luna, A. Acosta-Velázquez J. & Berlanga-Robles, C.A. (2008). On the reliability of the data of the extent of mangroves: A case study in Mexico. Ocean & Coastal Management, 51, pp. 342-351.

Ruiz-Luna, A., Cervantes E.A., & Berlanga-Robles C.A. (2010) Assessing distribution patterns, extent, and current Condition of northwest Mexico mangroves Wetlands, 30, pp. 717-723.

Sudhaka-Reddy, C. & Roy A. (2008). Assessment of three decade vegetation dynamics in mangroves of Godavari Delta, India using multi-temporal satellites data and SIG. Research Journal of Environmental Sciences, 2, 2, pp. 108-115.

Terchunian, A., Klemas, V., Segovia, A., Alvarez, A. Vasconez, B. & Guerrero, L., (1986) Mangrove mapping in Ecuador: the Impact of shrimp pond construction. Environmental management, 10, 3, pp. 345-350.

Tobey, J., Clay J. & Vergne, P. (1998) Maintaining Balance: The Economic, Environmental, and Social Impacts of Shrimp farming in Latin America, Coastal Management Report #2202, Coastal Resources Center at the University of Rhode Island, http://crc.uri.edu

2

Aquaculture and Environmental Protection in the Prioritary Mangrove Ecosystem of Baja California Peninsula

Magdalena Lagunas-Vazques,
Giovanni Malagrino and Alfredo Ortega-Rubio
Centro de Investigaciones Biológicas del Noroeste,
La Paz, Baja California Sur.
Mexico

1. Introduction

There are more than 123 coastal lagoons in the Mexican coastal zone covering an approximate area of 12,555 km². The length of these lagoons represents between 30 % and 35 % of the 11,543 km of the Mexican coast. Magdalena Bay, Mexico, is the largest bay in the Baja California peninsule (Fig. 1). The Bay is a lagoon system with three main areas, the northernmost called Laguna Santo Domingo, the central part Magdalena Bay, and the southernmost Almejas Bay. The lagoon system has a total length of 250 km, covering an area of 2,200 km² that includes 1,453 km² of the lagoon basin and 747 km² of mangrove forest, sand dunes, and wetlands (Malagrino, 2007).

Currently Magdalena Bay is very important for the economy of the state of Baja California Sur, 50 % of the artisan fisheries activities are established in this zone. To avoid conflicts between environmental conditions and commercial productive activities we studied and summarized the main biological, physical, chemical and socioeconomic aspects of Magdalena Bay, in order to determine where, and how, new clam culture projects must be established.

Aquaculture world production has maintained a sustained growth in several countries for the last 15 years, generating both positive and negative impacts, on social (Bayle, 1988; Primavera, 1991; Lebel *et al.*, 2002), economic (Kautsky *et al.*, 1997), and natural systems (Páez-Osuna, 2001; Macintosh, 1996). Mexico, being no exception, has developed these activities at similar rates going from 0 tons in 1984, to more than 62,000 tons in 2003 (SAGARPA-CONAPESCA, 2003).

Moreover, it is expected that aquaculture activities will increase explosively in the coming years. If aquaculture activities flourish, as it is foreseen in this region, there will be direct conflicts with artisan fishing activities, ecotourism and tourist activities foreseen to be carried out in the region, and with the conservation of the environment, including sand dunes and mangrove fragile ecosystems.

Magdalena Bay lagoon system is located on the occidental side of the state of Baja California Sur. This lagoon system is the most extensive and important of the whole peninsula and

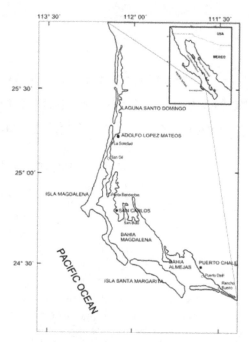

Fig. 1. Magdalena Bay location, Baja California Sur., México.

within Mexico. It is located between 24° 17′ and 25° 40′ N and 111° 30′ and 112° 15′ W. The system is made of wide areas of wetlands especially in Laguna Santo Domingo and in Almejas Bay.

Because of the physiography of the system it is regarded as a natural shelter for marine flora and fauna, and for small fishing boats. This zone is influenced by the California Current and by water that comes from the Equator, being a transition zone characterized by high productivity (Parrish et al., 1981); the climate is warm and dry, classified as a semiarid climate by Coppell system. The annual average temperature is of 20 °C, with a maximum of 41 °C in July-August, and a minimum of 4 °C in January-February. The mean total annual average temperature is of 125 mm (Rueda-Fernández, 1983). In the warm season water temperature column varies between 23 and 28 °C, while in the cold season it varies between 16 and 23.6 °C. Maximum salinity, ranging from 37.3 to 39.2 ups, is found in channels in the lagoon system, while minimum salinity, from 34.0 to 34.5 ups, is registered in channels connecting the system to the Pacific Ocean characterizing it as antiestuarine (Alvarez & Chee, 1975; Acosta-Ruiz & Lara-Lara, 1978). Tides are semi-diurnal mixed. Maximum and minimum dissolved oxygen level at the mouth of Magdalena Bay are of 6.85 and 3.68 mL/L respectively; concentration of chlorophyll a fluctuates from 1.2 to 5.1 mg/m^3; phosphates vary from 3.09 to 0.62 μm, and water velocity from 0.24 to 1 m/s (Rueda-Fernández, 1983).

2. Magdalena Bay Mangrove

Mangroves often provide a source of wood products, providing subsistence for local populations. However, logging is rarely the main cause of the loss of these trees. This is

Fig. 2. Satellite image of Magdalena Bay, Pacific Ocean, Mexico.

Fig. 3. Magdalena Bay Mangrove. (Photographer Magdalena Lagunas, 2011).

primarily due to competition for land for urban development, tourism, agriculture or construction of ponds for shrimp farming. The high rate of negative changes in the mangroves during the eighties in Asia, the Caribbean and Latin America has been caused mainly by the conversion of these areas for aquaculture and infrastructure, as many governments have opted for it with the intention of increasing security food, to stimulate national economies and improve living standards. According to FAO, in 1980 the mangroves covered a surface area of 19.8 million hectares of coastal areas of the world, for the year 2005, the same FAO report 15.2 million hectares, which means that in the past 20 years have been lost 23% of the global area. With the pressures and if the trend continues, we would be destroying one of representative ecosystems of global biodiversity (CONABIO, 2009). In Mexico, the annual loss rates calculated by comparing the mangrove areas are between 1 and 2.5%, depending on the method of analysis of the information used (INE, 2005).

Magdalena Bay is one of the largest lagoon system in Mexico. The dense mangroves of the bay represents the most extensive mangrove area of the Baja California peninsule (Enríquez-Andrade, 1998, Hastings & Fisher, 2001; Malagrino, 2007), 21, 116 has been holding 85% of mangrove state (Acosta-Velazquez & Vazquez-Lule, 2009), moreover, are of particular importance because of its isolation from other areas of its kind. Here, the mangroves are highly productive and structurally provide habitat, breeding sites and/or food for fish, crustaceans, molluscs, sea turtles (López-Mendilaharsu et al., 2005) and birds (Zárate, 2007). Particularly in these areas nesting a variety of both migratory and permanent residents seabirds (Hastings & Fisher, 2001).

Regarding to marine fauna, it has been reported 161 species of fish in the bay, belonging to 120 genera and 61 families and four species of sea turtles (Caretta caretta, Chelonia mydas, Dermochelys coriacea and Lepidochelys olivacea) listed as endangered in NOM-059-SEMARNAT-2001, gray whales (Eschrichtius robustus) under special protection and other marine mammals can also be found within the Bay (Tena, 2010). The marine flora of the lagoon system includes 279 species of macroalgae and 3 segrasses, for this reason it is considered as a bay with high vegetation species richness (Hernández-Carmona et al., 2007).

Magdalena Bay is defined as an area with a high level of ecological integrity, the National Commission for Biodiversity of Mexico, CONABIO recognizes the coastal area of Magdalena Bay as a priority region for conservation from the standpoint of terrestrial, marine, coastal and river basin hydrological as well as an Area of Importance for the Conservation of Birds (AICA). In recent national studies the Magdalena Bay, has been listed as a Site of Mangrove biological relevance and in need for ecological rehabilitation in the North Pacific Region, PN03 site identifier Baja California Sur, Magdalena Bay (Fig. 4), (CONABIO, 2009).

The most important plant community in the area is the mangrove. In Magdalena Bay exists three of the four Mexican mangrove species: red mangrove (Rhizophora mangle) which is endemic and is dominant in the area, associated with black mangrove (Avicennia germinans) and white mangrove (Laguncularia racemosa).

These species are listed under the category of special protection in the Mexican Official Standard NOM-059-SEMARNAT-2001. Mangroves are highly productive and structurally provide habitat and breeding and feeding sites for fish, crustaceans, molluscs, turtles and

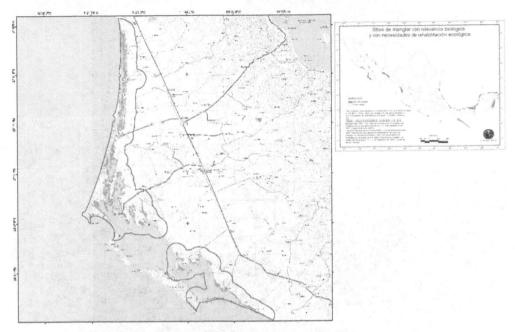

Fig. 4. Magdalena Bay mangrove (CONABIO, 2009).

birds. The mangroves of the lagoon system are the largest in the Baja California peninsule (Enriquez-Andrade *et al.*, 1998; Malagrino, 2007).

Recent studies on the coverage, distribution and structure of the mangroves of Magdalena Bay, indicates that the mangrove area is estimated above 17000 ha (Acosta-Velazquez & Ruiz-Luna, 2007). Mangrove class were subclassified into shrub, mixed and monospecific forest, based on field data (Acosta-Velazquez & Ruiz-Luna, 2007). The forest subclasses include *Rhizophora mangle* and *Laguncularia racemosa* species, being dominant the last species, with densities from 2339 to 5922 individuals/ha and basal area ranging from 20.6 to 58.5 m²/ha. Shrub mangroves includes also *Avicennia germinans*, whit densities up to 30000 individuals/ha. The mangrove area diminished more than 1500 ha between 1990 and 2005, with the shrub subclass as the most disturbed, while the monospecific forest displayed a significant increase (48%). The estimated annual mean deforestation rate was 0.55, but it was lower (0.15/year) if only the mangrove forests were included (Acosta-Velazquez & Ruiz-Luna, 2007).

Aquaculture and environmental impact on mangroves

The coastal zone bears most of the ecological consequences of aquaculture development. These include habitat loss/modification, excessive harvesting of wild seed/spawners and damage to bycatch, introductions of exotic species, escapes of cultured animals, spread of diseases, interactions with wild populations, misuse of chemicals and antibiotics, release of wastes, and dependence on wild fisheries.

Globally, more than a third of mangrove forests have disappeared in the last two decades, and shrimp culture is the major human activity accounting for 35% of such decline. This

Fig. 5. Magdalena Bay Mangrove, ecosystem Desert succession (Photographer Magdalena Lagunas, 2011).

transformation results in loss of essential ecosystem services generated by mangroves, including the provision of fish/crustacean nurseries, wildlife habitat, coastal protection, flood control, sediment trapping and water treatment. Fish pens and cages also degrade nearshore habitats through their physical installations on seagrass beds and sediment communities, or through deposits of uneaten feeds (Primavera, 2006).

The shrimp aquaculture ponds are located in the most biologically productive and undervalued in the world: marshes, mangrove forests and wetlands. It is clear that the mere physical presence of ponds for aquaculture production has an impact by hindering the continued natural flow between coastal environments. Mangrove conversion to shrimp ponds is the single major factor that has contributed to the negative press received by aquaculture. Southeast Asia has 35% of the world's 18 million ha of mangrove forests, but has also suffered from the highest rates of mangrove loss, e.g., 70–80% in the Philippines and Vietnam for the last 30 years. Around half of the 279,000 ha of Philippine mangroves lost from 1951 to 1988 were developed into culture ponds; 95% of Philippine brackish water ponds in 1952-1987 were derived from mangroves (Primavera, 2006)

Prevent environmental impacts on mangroves of Magdalena Bay

Although the ecological importance of Magdalena Bay is evident there are currently no state or federal programs that regulate the lagoon complex ecological system or the area surrounding it. The gradual increase in population and alternative activities such as ecotourism will only increase the ecological pressures on the ecosystem.

An important contribution to the proper use of this lagoon system is precisely to obtain adequate rates to define the proper establishment of suitable sites for different types of aquaculture farming.

In order to consider the suitability of each potential site for the clam culture, a mathematical index was utilized. This index helped determine the ranks of suitability of each site in regards to its possible use for marine cultures. Based on the results obtained, seven sites were identified as adequate for clam culture. The one that presents the best conditions is estuary San Buto. The Methodology and Results of this work can be used in all Coastal Zone with Aquaculture potentialities.

Mangroves and aquaculture are not necessarily incompatible. For example, seaweeds, bivalves and fish (seabass, grouper) in cages can be grown in mangrove waterways. Such mangrove-friendly aquaculture technologies are amenable to small-scale, family-based operations and can be adopted in mangrove conservation and restoration sites. Brackish water culture ponds may not necessarily preclude the presence of mangroves. Dikes and tidal flats fronting early Indonesian tambak were planted with mangroves to provide firewood, fertilizers and protection from wave action. Present-day versions of integrated forestry-fisheries-aquaculture can be found in the traditional geiwai ponds in Hong Kong, mangrove-shrimp ponds in Vietnam, aquasilviculture in the Philippines, and silvofisheries in Indonesia. Alternatively, mangroves adjacent to intensive ponds can be used to process nutrients in pond effluents (Primavera, 2006).

Fig. 6. Suspension system in aquatic farming Baja California peninsula.

2.1 Methodology

From 2001 to 2003 the main characteristics of the area were determined: climate, soil, geology, orography, morphology, and hydrology; after analyzing the bibliography and the data sets of the meteorological stations of the region as well as official charts, field stages of work were developed to corroborate the information.

Marine and costal characteristic, including tide effects, morphology of the coastal zone, and accessibility for marine water intake and waste water treatment and disposal were obtained through the analysis of satellite images and field stays of work. The availability of services for each potential clam culture zone, including roads (paved and not paved), electricity, phone and internet availability, human populations and potential workers, were also established by field surveys.

In order to assess the suitability of each potential site for sustainable clam culture activities inside Magdalena Bay, we applied the modified index of Lagunas & Ortega (Lagunas, 2000):

$$CCS = \frac{ACS\,(0.1) \;+\; NMST\,(0.3) \;+\; AVD\,(0.15) \;+\; MCAS\,(0.15) \;+\; AAPD\,((0.3) \;+\; MSK\,(0.1) \times 100}{2.45}$$

Where:

CCS = Clam Culture Suitability

ACS = Accessibility

NMST = Number of Months with Suitable Temperature for the Clam Culture

AVD = Average Depth

MCAS = Marine Current Average Speed

AAPD = Annual Average Phytoplankton Density

MSK = Marine Substrate Kind

The index of Lagunas & Ortega is the result of empirical field studies developed by the authors of this paper and by the careful assessment of the ecological, environmental, socioeconomic and facilities characteristics of the places where the optimal clam aquaculture activities are currently developed. In order to standardize the values obtained by this index, it is divided by the empirical value of 2.45. This way values obtained range from 0 to 100. The rank values are shown in Table 1.

3. Conclusions

After the bibliographic revision and the analysis of satellite images, which were digitalized in a GIS in which we included climate, vegetation, soil, geological, and geomorphological characteristics, 7 potential places were identified where clam culture activities could be performed with the lesser impacts and with more success probability; the sites selected were Santo Domingo, Adolfo Lopez, Estero San Buto, Estero Chisguete, Punta Cayuco, Puerto Chale, and Rancho Bueno, which are shown in Figure 7.

Variable	Characteristics	Rank value
ACS	Site without roads available and without sea connection Site only with sea connection Site only with road connection Site with roads available and with sea connection	0 1 2 3
NMST	Less than 5 months with suitable temperature Between 6 to 9 months with suitable temperature More than 10 months with suitable temperature	0 1 2
AVD	From 1 to 5 m in average From 6 to 10 m in average More than 10 m	1 2 3
MCAS	Annual average more than 31 m/s Annual average between 21 to 30 m/s Annual average between 11 to 20 m/s Annual average between 1 to 10 m/s	0 1 2 3
AAPD	From 67% to 100% of water turbidity From 34% to 66% of water turbidity From 11% to 33% of water turbidity From 1 % to 10% of water turbidity	0 1 2 3
MSK	Rocky Clay-slime Sandy	0 1 2

Table 1. Rank value used to assess clam aquaculture suitability.

After the 7 sites were selected, we performed one week field stays of work in each of them. The CCS value for each site was obtained after we analyzed all the specific characteristics of the environmental and socioeconomic features considered in the formula for each place (Table 2). As we can see the best site to develop clam culture is located in Estero San Buto where optimal conditions were found with a CCS value of 90.

According to the characterization of this site we recommend that the species to be cultured are the Catarina scallop (*Argopecten ventricosus*), the Lion paw scallop (*Nodipecten subnodosus*), the Pen shell (*Atrina maura*), and the Chocolata clam (*Megapitaria squalida*).

We recommend to establish polyculture facilities for these local species, without intensive culture activities, thus avoiding the nourishment excess and the pollution caused by the own detritus of the cultured individuals (Fig. 4). There exist 150 ha in Estero San Buto suitable to establish this recommended polyculture, being possible to hire workers in the locality for it; an average annual harvest of 2,312 tons of the different species is estimated.

Knowing the precise biological, physicochemical and social environment, we can determine the best species to cultivate, the recommended total area to use, and the methodology to be employed to produce the lesser environmental impacts and to obtain the maximum profitability. Our methodology could be used not only to select appropriate sites for clam culture but also to assess the suitability, in a quick and accurate way, of any other aquaculture activity in coastal zones.

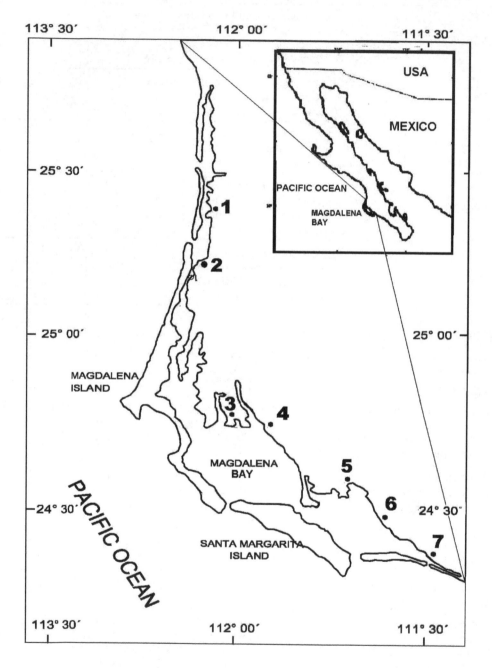

Fig. 7. Magdalena Bay location and selected clam culture sites. Sites are described in Table 2.

SITE	ACS Accessibility	NST Number of Months With Suitable Temperature for the Clam Culture	AVD Average depth	MCAS Marine Current Average Speed	AAPD Annual Average Phytoplankton Density	MSK Marine Substrate Kind	CCS Clam Culture Suitability
1 Santo Domingo	1	2	1	1	1	1	57
2 Adolfo Lopez	3	2	1	1	1	1	65
3 Estero San Buto	2	2	2	2	2	2	90
4 Estero Chisguete	2	2	3	1	1	1	73
5 Punta Cayuco	1	2	1	1	2	1	69
6 Puerto Chale	3	2	1	1	1	2	65
7 Rancho Bueno	2	2	1	1	1	2	65

Table 2. Main characteristics and CCS value obtained for each selected site.

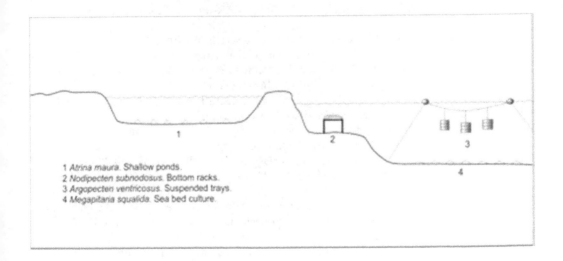

1 *Atrina maura*. Shallow ponds.
2 *Nodipecten subnodosus*. Bottom racks.
3 *Argopecten ventricosus*. Suspended trays.
4 *Megapitaria squalida*. Sea bed culture.

Fig. 8. Recommended polyculture.

Mangrove-friendly aquaculture technologies are amenable to small-scale, family-based operations and can be adopted in mangrove conservation and restoration sites.

Make mangrove reforestation, to consider forestry to aquaculture (eg. the traditional geiwai ponds in Hong Kong, mangrove shrimp ponds in Vietnam on aquasilviculture in the Philippines and in Indonesia silvofisheries). On the other hand, mangroves adjacent to the intensive ponds can be used to process the nutrients in pond effluents.

Magdalena Bay mangroves seems to be exposed to relatively low human impact compared whit other Mexican system of mangrove. Effective conservation programs and sustainable management plans will be required to preserve this important mangrove ecosystem, mainly related to aquaculture activities, based in scientific knowledge such as the study here presented.

4. Acknowledgments

This study was supported by the Centro de Investigaciones Biológicas del Noroeste, by the Project SEMARNAT- CONACyT: 2008-CO1-107923, and by the Universidad Autónoma de Baja California Sur, PROMEP, SEP. We thank the two anonymous reviewers whose pertinent opinions were important to write a better version of this manuscript.

5. References

Acosta-Ruiz, M. and Lara-Lara, J. (1978). Resultados físico-químicos en un estudio de variación diurna en el área central de Bahía Magdalena, B. C. S. *Ciencias Marinas*, Vol.5, No.1, pp.37-46.

Acosta-Velazquez J. & Ruiz-Luna A. (2007). Variación en la cobertura, distribución y estructura de los manglares del complejo lagunar Bahía Magdalena-Bahía Almejas (1900-2005). En Estudios Ecológicos en Bahía Magdalena. Editores: Funes R.R., Gómez G.J. & R. Palomares. CIBNOR-CICIMAR-IPN. La Paz, BCS. México. Pp 113-126.

Acosta-Velázquez, J. & A. D. Vázquez-Lule. Caracterización del sitio de manglar Bahía Magdalena, en Comisión Nacional para el Conocimiento y Uso de la Biodiversidad CONABIO. (2009). Sitios de manglar con relevancia biológica y con necesidades de rehabilitación ecológica. CONABIO, México, D.F.

Alvarez, S., Galindo, L., and Chee, A. (1975). Características hidroquímicas de Bahía Magdalena, B.C. S. *Ciencias Marinas*, Vol.2, No.2, pp. 94-110.

Bailey, C. (1988). The social consequences of tropical shrimp mariculture development. *Ocean and Shoreline Management*, Vol.11, No.1, pp. 31-44.

Comisión Nacional para el Conocimiento y Uso de la Biodiversidad CONABIO. (2009). Sitios de manglar con relevancia biológica y con necesidades de rehabilitación ecológica. CONABIO, México D.F.

CONABIO-SEMAR (March 2011). Mangrove sites with relevant ecological. Available from http://www.conabio.gob.mx/conocimiento/manglares/doctos/sitios.html

Enríquez-Andrade, R. R. (1998). Resumen Ejecutivo del Programa de acciones y áreas prioritarias para la conservación en la Península de Baja California 1998-2007. Pronatura. Península de Baja California. Ensenada, B. C.

Hastings, R. M. & D. W. Fisher. (2001). Management priorities for Magdalena Bay, Baja California, México. *Journal of Coastal Conservation.* No.7: pp 193-202. http://www.sagarpa.gob.mx/conapesca/planeacion/anuario/anuario2003.pdf

INE. (2005). Evaluación Preliminar de las Tasas de Pérdida de Superficie de Manglar en México. Dirección General de Ordenamiento Ecologico y Conservación de Ecosistemas. SEMARNAT-INE. Documento técnico. 21 pp

Kautsky, N., Berg, H., Folke, C., Larson, J. and Troell, M. (1997). Ecological footprint for assessment of resource use and development limitations in shrimp and tilapia aquaculture. *Aquaculture Research,* Vol.28, No.10, 753-763.

Lagunas, M. (2000) Identificación del uso potencial de áreas costeras en B.C.S., para desarrollar la acuicultura de camarón con base en las condiciones del ambiente natural. Thesis, Bachelor Degree. Universidad Autónoma de Baja California Sur. La Paz, B. C. S., Mexico.

Lebel, L., Hoang-Tri, N., Saengnoree, A., Pasong, S., Butama, B. and Kim-Thoa, L. (2002). Industrial transformation and shrimp aquaculture in Thailand and Vietnam: Pathaways to ecological, social, and economical sustainability? *AMBIO,* Vol.31, No.4, pp. 311-323.

López-Mendilaharsu, M., S. C. Gardner, R. Riosmena-Rodrìguez, & J. A. Seminoff. (2005). Identifying critical foraging habitats of the green turtle (Chelonia mydas) along the Pacific Coast of the Baja California Península, México. *Aquatic Conservation: Marine and Freshwater Ecosystems.* Vol. 15 No. 3: pp 259-269.

Macintosh, D. J. (1996). Mangroves and coastal aquaculture: doing something positive for the environment. *Aquaculture Asia,* Vol. 2, No.2, pp. 3-10.

Malagrino Lumare, G. (2007). Manejo de zona costera en Bahía Magdalena, B. C. S.: Cultivode Organismos Marinos. Ph.D Thesis. CIBNOR, S. C., La Paz, Mexico.78 pp.

Páez-Osuna, F. (2001). Impacto ambiental y desarrollo sustentable de la camaronicultura. *Ciencia,* Vol.52, No.12, pp.15-24.

Parrish, R. H., Nelson, C. S. and Bakun, A. (1981). Transport mechanisms and reproductive success of fishes in California Current. *Biological Oceanography,* Vol.1, No.2, pp. 175-203.

Primavera, (2006). Overcoming the impacts of aquaculture on the coastal zone. *Ocean & Coastal Management,* No.49 (2006) pp. 531–545

Primavera, J.H. (1991). Intensive prawn farming in the Philippines: ecological, social, and economic implications. *AMBIO,* Vol.20, No.1, 28-33.

Rodríguez. (2007). Flora Marina del sistema lagunar de Bahía Magdalena-Bahía Almejas. En Estudios Ecológicos en Bahía Magdalena. Editores: Funes R.R., Gómez G.J. & R. Palomares. CIBNOR-CICIMAR-IPN. La Paz, BCS. México. Pp 127-126-141.

Rueda-Fernández, S. (1983). La precipitación como indicador de la variación climática de la península de Baja California y su relación dendrocronológica. Master Thesis. Centro Interdisciplinario de Ciencias Marinas. La Paz, B.C.S., Mexico.

SAGARPA-CONAPESCA (2003). Anuario Estadístico de Pesca 2003. SAGARPA, Mexico. Available from

Tena G. A. (2010). Determinación de Áreas Prioritarias para la Conservación de la Biodiversidad en la Zona Costera e Islas de Bahía Magdalena, B. C. S., México.

Master of Science Thesis. Cibnor S.C. La Paz. México. 110 pp. Hernández-Carmona G., Serviere-Zaragoza E., Riosmena-Rodriguez R., I Sánchez-

Zárate Ovando, M. B. (2007). Ecología y conservación de las aves acuáticas del complejo lagunar Bahía Magdalena-Almejas, B. C. S., México. Ph.D Thesis. CIBNOR, S. C., La Paz, Mexico.

Mangrove Revegetation Potentials of Brackish-Water Pond Areas in the Philippines

Maricar S. Samson[1,2] and Rene N. Rollon[3]
[1]*Br. Alfred Shields FSC Marine Station, De La Salle University, Manila*
[2]*School of Environmental Science and Management,*
University of the Philippines Los Baños
[3]*Institute of Environmental Science and Meteorology,*
University of the Philippines Diliman
Philippines

1. Introduction

The Philippines is one of the countries with the most number of true – mangrove species (about 42 species, 18 families, **Table 1**) (Primavera, 2004; Spalding et al 2010; Polidoro et al 2010). However Philippine mangrove forests suffered greatly from anthropogenic activities, i.e. cutting for firewood and charcoal, siltation caused by upland deforestation, and conversion of mangrove areas to shrimp ponds, fishponds and salt ponds (Primavera 1991, 1995, 2000; Field, 1998; FAO, 2003, 2007). From 1918 (~450,000) to 1998 (112,400), mangrove cover declined by more than 75% (**Figure 1**). In 2007, the remaining mangrove areas in the Philippines was estimated at 289,350 hectares (DENR-NAMRIA 2007), a value which is 61% (176,950) higher than 1998 estimate. However, most of these are estimates based on satellite images that need to be validated on field.

The typical historical zonation of mangrove species in the Philippines follows that described by Duke et al in 1998 for mangroves found along Daintee River in Australia. Species with pneumatophores are commonly found at the low-intertidal; prop- and knee roots species are in the mid-intertidal; and buttress or plank root species are at the high intertidal area (**Figure 2a**). However due to the aforementioned large scale conversion to aquaculture ponds, the mangrove communities at the middle zone were diminished (**Figure 2b**).

Of the total mangrove areas that were deforested, sixty-eight percent were converted to brackish-water ponds (Primavera, 1995, 2000). One of the legal instrument of operation of brackish-water ponds in the Philippines is the Fishpond Lease Agreement (FLA) that is being granted by the Philippine Bureau of Fisheries and Aquatic Resources (BFAR) under the Department of Agriculture (DA). In 2007, there were 59,923 hectares of potential brackish-water ponds with FLA belonging to 4,386 registered operators (BFAR, n.d.). However, the license agreement of almost 65% (39,152 ha) of these brackish-water ponds with FLA are already expired and as of the list posted in December 2010 had not been renewed. **Table 2** presents the details of the top 10 provinces in terms of the area with expired FLA licenses in the Philippines.

FAMILY[1]	SPECIES[1,2]
ACANTHACEAE	*Acanthus ebracteatus*
	Acanthus ilicifolius
ARECACEAE	*Nypa fruticans*
AVICENNIACEAE	*Avicennia alba*
	Avicennia marina
	Avicennia officinalis
	Avicennia rumphiana
BIGNONIACEAE	*Dolichandrone spathacea*
BOMBACACEAE	*Camptostemon philippinense*
	Camptostemon schultzii
CAESALPINIACEAE	*Cynometra iripa*
COMBRETACEAE	*Lumnitzera littorea*
	Lumnitzera racemosa
EBENACEAE	*Excoecaria agallocha*
LYTHRACEAE	*Pemphis acidula*
MELIACEAE	*Xylocarpus granatum*
	Xylocarpus moluccensis
MYRSINACEAE	*Aegiceras corniculatum*
	Aegiceras floridum
MYRTACEAE	*Osbornia octodonta*
PLUMBAGINACEAE	*Aegialitis annulata*
PTERIDACEAE	*Acrostichum aureum*
	Acrostichum speciosum
RHIZOPHORACEAE	*Bruguiera cylindrica*
	Bruguiera exaristata
	Bruguiera gymnorhiza
	Bruguiera hainesii
	Bruguiera parviflora
	Bruguiera sexangula
	Ceriops decandra
	Ceriops tagal
	Kandelia obovata
	Rhizophora apiculata
	Rhizophora mucronata
	Rhizophora stylosa
	Rhizophora x lamarckii
RUBIACEAE	*Scyphiphora hydrophylacea*
SONNERATIACEAE	*Sonneratia alba*
	Sonneratia caseolaris
	Sonneratia ovata
	Sonneratia x gulngai
STERCULIACEAE	*Heritiera littoralis*

Table 1. List of true mangrove species in the Philippines (Spalding et al, 2010[1]; Polidoro et al, 2010[2]).

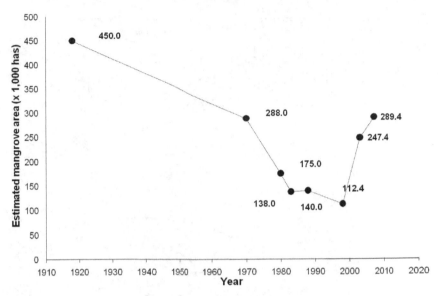

Fig. 1. Estimated extent of mangrove areas in the Philippines from 1918 (around 450,000 hectares) to 2007 (around 289,350 hectares).

PROVINCE	TOTAL FLA AREAS	NO. OF LESSEE	AVERAGE AREA GRANTED	TOTAL AREA WITH EXPIRED FLA	% EXPIRED
Antique	150.65	5	30.13	150.65	100%
Maguindanao	125.36	6	20.89	125.36	100%
Lanao del Norte	1141.11	26	43.89	1097.07	96%
Palawan	1422.96	25	56.92	1356.19	95%
Basilan	854.04	43	19.86	804.80	94%
Sulu	172.80	2	86.40	159.00	92%
Davao Oriental	369.58	17	21.74	334.42	90%
Sultan Kudarat	291.73	11	26.52	262.83	90%
Northern Samar	1010.24	34	29.71	893.87	88%
Camarines Sur	1181.93	75	15.76	1017.92	86%

Table 2. Top 10 provinces in terms of the area with expired FLA licenses in the Philippines.

It has long been noted by various authors that most of the brackish-water ponds in the Philippines are either idle, abandoned, underutilized and in an unproductive state (Primavera and Agbayani 1997; Primavera 2000; Yap 2007; Samson and Rollon 2008; Primavera et al, 2012). There had been efforts to promote the reversion of these ponds to mangrove areas however the implementing rules and regulations had been unclear, if not nonexistent. In most cases, holders of FLA certificates are unwilling to yield the pond area for mangrove restoration (Yap, 2007; Samson and Rollon 2008). These underutilized ponds

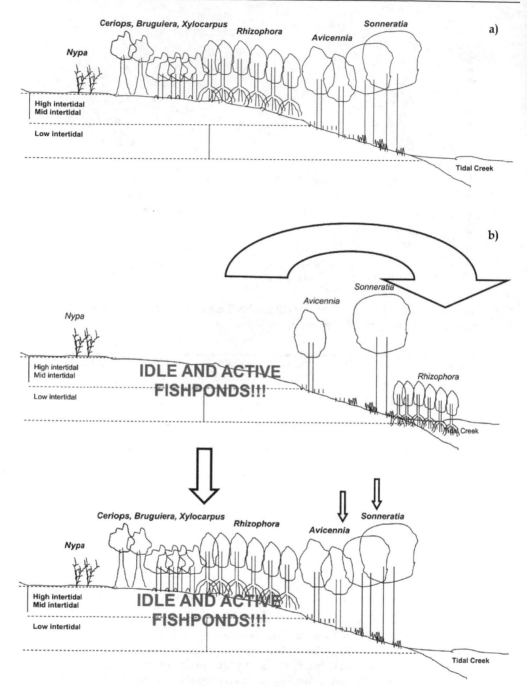

Fig. 2. Illustration of the a) historical and b) present condition and c) best management options for the mangroves and brackish-water pond areas in the Philippines.

perpetuates the loss of goods and services that mangrove areas could provide. The continued abandonment of these areas increases the vulnerability of coastal communities to the ancillary impacts of climate change such as increase in sea level (Alongi, 2002; Gilman et al, 2006; Gilman, 2008), tsunami (Alongi 2008; Dahdouh-Guebas, 2005; Vermaat & Thampanya, 2006), wave impact due to increased typhoon strength and frequency and coastal erosion (UNEP-WCMC, 2006; Primavera et al 2012).

If massive loss of mangrove areas in the Philippines could be attributed to aquaculture development, logically therefore, restoration of idle and underproductive brackish-water ponds at least to its ecologically productive state, should be the focus of management efforts (Primavera, 2006; Samson & Rollon, 2008; Primavera & Esteban, 2008; Primavera et al, 2012). This option will greatly enhance the ecological success of current efforts by 1) promoting healthy growth patterns of planted species, and 2) stop the afforestation of adjacent habitats (i.e. seagrass bed and mudflat area, **Plate 1**). These practices are widespread in the Philippines where the growth of species in afforested sites performed dismally as compared to those planted in natural mangrove forests (Samson and Rollon, 2008; **Figure 3**). Though the revegetation of idle and unproductive ponds may present a multitude of ecological, political and institutional challenges (Primavera, 2000; Samson and Rollon, 2008) to become feasible, conscious effort to move towards this objective must be prioritized. **Table 3** lists some factors that needs to be considered before deciding the reversion of idle or disused ponds to mangrove areas.

Plate 1. Some examples of the well-meaning planting initiatives but may be less successful in terms of ecological restoration in Talibon, Bohol, Philippines where mangroves were planted on seagrass and mudflat areas.

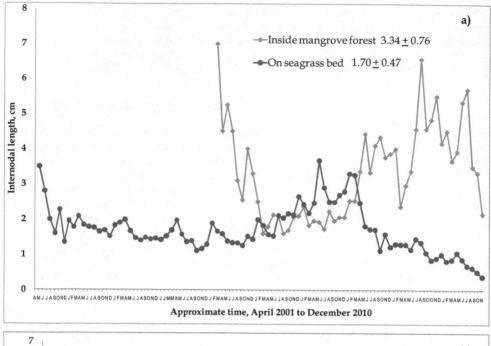

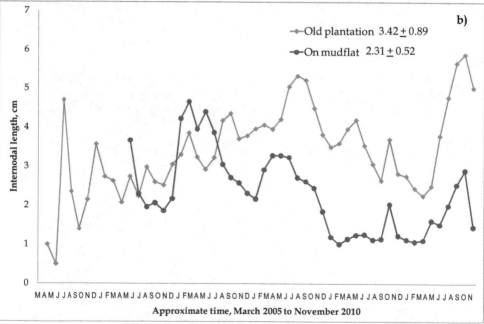

Fig. 3. A comparison of the internodal growth patterns of *Rhizophora stylosa* planted in different mangrove planting sites in Talibon, Bohol, Philippines: a) Cataban Island and b) Barangay San Francisco.

	CONSIDERATIONS
Ecological	If farm was constructed in an inappropriate site such that: Farm density is already too high Occurrence of diseases There is insufficient water supply Too much rainfall; Unsuitable soils Frequently hit by typhoons (Stevenson et al 1999)
	Poor productivity level due to high acidity and poor shrimp survival (Stevenson et al 1999)
	Further degradation of the ecosystem – i.e. acidification, soil erosion – if some rehabilitation activity is not undertaken (Stevenson et al 1999)
	Arrest surface erosion and subsidence and compaction of soil profile within and in the adjacent environment (Burbridge and Hellin, 2002)
	Tidal hydrology can still be restored (Lewis 2005)
	Availability of water borne seedlings and propagules from neighboring mangrove communities (Lewis 2005)
	Possibility of occurrence of natural process of secondary succession (Lewis 2005)
Economic	Non – sustainable and unproductive pond operation (Stevenson et al 1999)
	If degradation not arrested, repair may become progressively more expensive and difficult – rehabilitation costs would be balanced by costs avoided (Stevenson et al 1999)
Social	Willingness and cooperation of stakeholders (Stevenson et al 1999)
	Provide additional protection from strong waves; if infrastructures for protection against strong waves and typhoons are more costly than revegetation efforts (Stevenson et al 1999; Primavera 2000)
Institutional	Expired FLAs
	Existing comprehensive land – use plan
	Maintenance of mangrove greenbelt as required by law – Fishery Reform Code (Primavera, 1995)

Table 3. Factors to consider before deciding the reversion of idle or underproductive ponds to mangrove areas.

The issues surrounding the decline of mangrove areas in the Philippines and the proliferation of idle and underutilized brackish-water ponds in terms of area covered are inextricably linked and may be addressed in a more integrated and adaptive approach (**Figure 4**). **Figure 2c** illustrates what may be the best management options for mangroves and brackish-water ponds in the Philippines. As cited by Primavera and Esteban (2008), Saenger et al (1986) recommended the 4:1 mangrove to pond ratio to sustain the ecological function of the mangrove ecosystem. However this recommendation poses a great challenge to the Philippine government as the basic information on the present state of ownership and operation of these brackish-water ponds are lacking.

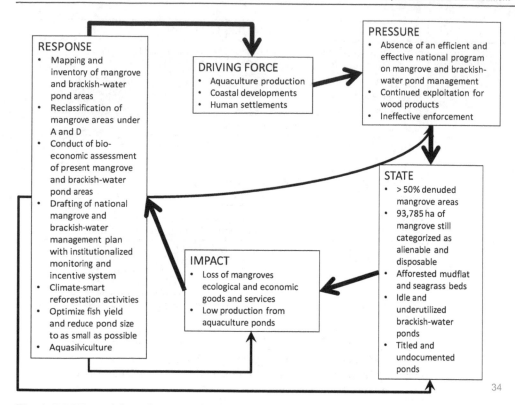

Fig. 4. DPSIR model on the state of mangrove forests and brackish-water ponds in the Philippines.

2. Mangrove-pond ratio of selected aquaculture production centers in the Philippines

As presented in the previous section, brackish-water ponds now occupy a large part of the natural mangrove areas in the Philippines. In the provinces with the largest area of brackish-water ponds in the Philippines, mangrove loss is more than 75% of the natural forest. The mangrove-pond ratio in these areas ranges from 1:2 to 1:1,586 (**Table 4**). However 44 to 99% percent of these brackish-water pond areas are not covered by FLA (**Table 5). Figure 5** presents the potential mangrove extent and hectarage of brackish-water pond area, with or without FLA in selected sites of the country. On the average, around 50% of the historical mangrove areas in these selected sites were converted to brackish-water ponds with Iloilo City having the highest percentage of converted area (90%). However, as in the situation in many provinces around the country, not all of the brackish-water ponds have FLAs, on the average only 40% are under the 25-years lease agreement with the government (**Table 6**). Worse, only 60% of these FLAs are still active. What could have happened to the other mangrove areas that were converted to brackish-water ponds? One possible answer to this is that the other areas have land titles or undocumented as in the case of the 72 has (92%) of ponds in Lian, Batangas.

Province	Estimated mangrove extent (ha, 2007)	Brackish water pond area (ha, 1994)	Mangrove to pond ratio
Former Zamboanga del Sur	19,830	32,992	1:2
Pampanga	220	22,231	1:101
Bulacan	10	15,861	1:1586
Negros Occidental	2,790	9,796	1:3
Capiz	2,810	8,404	1:3
Pangasinan	170	7,026	1:41
Iloilo	1,350	5,504	1:4
Masbate	2,440	4,974	1:2
Quezon	14,540	4,876	3:1
Aklan	360	4,653	1:12

Table 4. Estimated mangrove extent, brackish-water pond area and mangrove to pond ratio in the top 10 provinces with the largest hectarage of brackish-water ponds in the Philippines.

Province	Brackish water pond area (ha, 1994)	BW ponds with active FLAs (ha, 2010)	BW pond not covered by FLAs (ha, 2010)	% BW pond with no FLA
Former Zamboanga del Sur	32,992	1,700.81	31,291.19	94.8
Pampanga	22,231	79.27	22,151.73	99.6
Bulacan	15,861	70.00	15,791.00	99.6
Negros Occidental	9,796	1,514.52	8,281.48	84.5
Capiz	8,404	791.56	7,612.44	90.6
Pangasinan	7,026	1,132.02	5,893.98	83.9
Iloilo	5,504	3,067.12	2,436.88	44.3
Masbate	4,974	929.35	4,044.65	81.3
Quezon	4,876	1,576.41	3,299.59	67.7
Aklan	4,653	1,599.76	3,053.24	65.6

Table 5. The estimated brackish-water pond area, ponds with and not covered by FLA as of 2010 in the top 10 provinces with the largest hectarage of brackish-water ponds in the Philippines.

PROVINCE	MUNICIPALITY	POND AREA (ha)	FLA AREA (ha, 2010)
Batangas	Calatagan	299.43	111.21
	Lian	78.64	6.60
Quezon	Calauag	533.76	479.75
Sorsogon	Sorsogon City	318.81	137.06
	Prieto Diaz	463.48	299.40
Iloilo	Iloilo City	871.92	2.54
Former Zamboanga del Sur	Aurora	3,955.86	524.53
	Kabasalan	992.82	588.33

Table 6. Mangrove extent, brackish-water pond and FLA areas in selected sites of the Philippines.

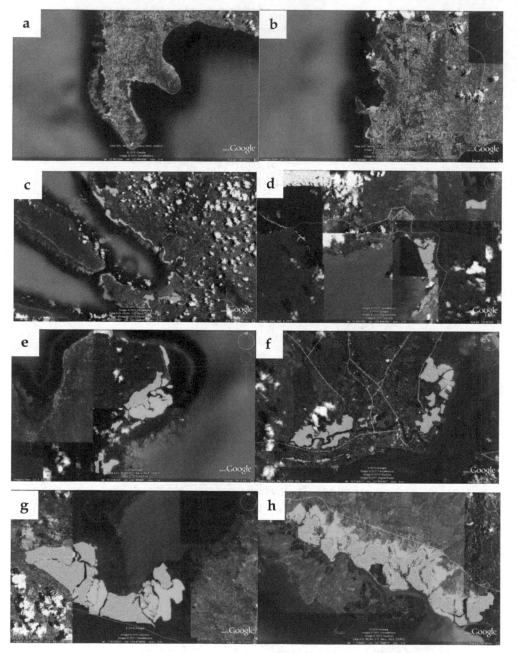

Fig. 5. Digitized Google Earth images of potential mangrove (green) and brackish-water ponds areas (dark yellow) in selected sites of the Philippines, a) Calatagan and b) Lian, Batangas, c) Calauag, Quezon, d) Sorsogon City, e) Prieto Diaz, Sorsogon, f) Iloilo City, g) Aurora, Zamboanga del Sur, and h) Kabasalan, Zamboanga Sibugay.

3. Pond area-fish yield in the past decade

The contribution of the aquaculture sector is still the highest (from 38 to 49%) in terms of the volume of production in the last decade (**Figure 6**). Almost fifty percent of the total fisheries production in 2010 comes from this sector. Brackish-water pond aquaculture is still the subsector with the highest percentage of production in terms of volume (**Figure 7**). However

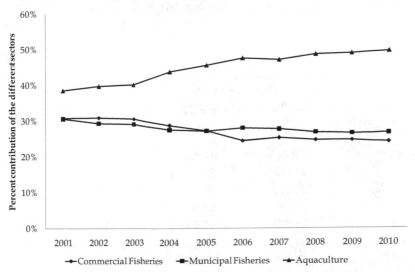

Fig. 6. Percent contribution of the different sectors of fisheries production in the Philippines from 2001 to 2010.

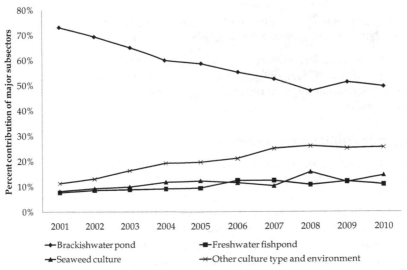

Fig. 7. Percent contribution of the different subsectors of aquaculture production in the Philippines from 2001 to 2010.

a steady decline of production from this sector is apparent from 2001 to 2010 and this may be due to the under productivity of almost 40% of brackish-water ponds in the country. Included in these areas are three of the provinces with the largest hectarage of brackish-water ponds, namely former Zamboanga del Sur, Bulacan and Aklan (**Table 7**).

Province	Volume of production (mt, 2010)		% contribution of BW ponds to total aquaculture production	Estimated annual production from BW ponds per hectare
	Aquaculture, total	Brackish-water pond		
Former Zamboanga del Sur	187,718	8,947.2	5%	0.27
Pampanga	148,603	47,400.1	32%	2.13
Bulacan	41,182	33,459.6	81%	2.11
Negros Occidental	23,200	20,411.0	88%	2.08
Capiz	40,216	26,340.3	65%	3.13
Pangasinan	85,523	15,645.2	18%	2.23
Iloilo	23,235	21,111.4	91%	3.84
Masbate	2,499	1,059.8	42%	0.21
Quezon	55,688	11,863.7	21%	2.43
Aklan	8,987	8,196.3	91%	1.76

Table 7. Volume of total aquaculture and brackish-water pond production in the top 10 provinces with the largest hectarage of brackish-water ponds in the Philippines.

In terms of the physical area utilized, brackish-water aquaculture is the biggest sub-sector in Philippine aquaculture (Cruz, 1997), however, its contribution to total fisheries production may not be proportionate with its physical magnitude in terms of area covered (~169,000 ha in 1995, Yap, 2007).

4. Management options for brackish-water pond areas in the Philippines

Brackish-water ponds in the Philippines may be classified into two: 1) those with valid legal instrument of ownership and operation, 2) those that do not have legal instrument of operation and undocumented. The first classification are of two types, those with FLAs and those with land titles. As presented in the previous section, 40 to 50% of these ponds, are now left idle or underproductive. A number of authors discussed several management strategies to address this problem (Primavera, 2000, 2006; Yap, 2007; Samson and Rollon, 2008; Primavera and Esteban, 2008; Primavera et al, 2012). Samson and Rollon in 2008 proposed a possible decision tree of options for idle and active brackish-water ponds in the Philippines. These options are specifically for those with FLAs.

The present paper is an attempt to extend the options presented by Samson and Rollon (2008) to include brackish-water ponds covered by land titles and those that are undocumented (**Table 8**).

PONDS WITH LEGAL INSTRUMENTS		PONDS WITHOUT EXISTING LEGAL INSTRUMENT OF OWNERSHIP
With active FLA	Titled ponds	
1. For ponds with active FLAs and the overall fish yield is optimal, apply semi-intensive aquaculture. (modified from Samson and Rollon, 2008). 2. Optimize fish yield and reduce pond size to as small as possible (Samson and Rollon, 2008). Follow the 4:1 mangrove-pond area ratio to maintain ecological health (as cited in Primavera, 2000).	1. Aquasilviculture (Primavera, 2000) 2. Optimize fish yield and reduce pond size to as small as possible (Samson and Rollon, 2008); revert unutilized areas to mangrove forest	1. If pond existence is necessary based on bioeconomic analysis, reapply FLA and optimize fish yield (Samson and Rollon, 2008). 2. Revert to mangrove areas: a. Natural revegetation b. Assisted planting

Table 8. List of management options for brackish-water ponds with and without legal instruments (with FLA, land titles) in the Philippines.

A total of six rational management strategies are being proposed for the utilization of brackish-water ponds in the Philippines. The options are specific for the current state of ownership of the ponds.

4.1 For ponds with active FLAs and the overall fish yield is optimal, apply semi-intensive aquaculture (modified from Samson and Rollon, 2008)

For ponds with existing FLA, the most rational objective of management will be semi-intensive aquaculture production to optimize the use of leased areas (Janssen and Padilla, 1999). In 2007, there were around 59, 923 hectares of potential brackish-water ponds listed in the website of the Bureau of Fisheries and Aquatic Resources. However, a closer look at the list revealed that only 44% of these have active FLAs, the rest of the lease are expired. This option may be of importance to the provinces of Quezon, Zamboanga del Sur, Iloilo, Occidental Mindoro, Negros Occidental, Samar, Masbate, Bohol, Zamboanga City and Capiz where brackish-water pond areas with FLAs are relatively extensive. Applying this intervention may yield a net income of US$680 million (at US$1 = Php 42.61) conservatively in 10 years.

4.2 For ponds with active FLAs and the fish yield is not anymore financially sustainable, production needs to be optimized and pond size may be reduced to as small as possible whereby a 4:1 mangrove-pond area ratio may be followed to restore and maintain the ecological health of the system (Primavera, 2000; Samson and Rollon, 2008; Primavera and Esteban, 2008)

This option particularly addresses brackish-water ponds with FLAs but are not operating sustainably. Technical assistance from concerned agencies (i.e. BFAR) must be sought to optimize the production of these ponds. However, it may also be that the reason for the under productivity of these ponds is that the area may not anymore be suitable for production hence the strategy on reducing the pond size and following the 4:1 mangrove-pond area ratio is being put forward (Primavera, 2006). This option on the reversion of some ponds with existing FLA to mangroves will require joint efforts from concerned agencies (i.e. Department of Environment and Natural Resources (DENR), BFAR, Department of Interior and Local Government (DILG)) and the institutionalization of strategies on how the pond operators will be convinced to reforest their underutilized ponds. Strategies may be in the form of incentives like granting of awards for operators with the most environment friendly operations. Another strategy may be is to highlight the importance of these reverted areas in bringing back mangrove goods and services which may become a sustainable source of income in the form of mangrove associated fisheries. The role of these former mangrove areas in mitigating the impacts of climate change such as sea level rise and increased storminess may also convince pond owners to reforest their idle and unproductive ponds. A more political-institutional approach will be to strengthen the monitoring of pond operations such that the 5 years limit for unproductive ponds will be imposed and those ponds will be reverted back to DENR. For the government, specifically BFAR, this is an opportunity to improve its system of monitoring and evaluation of existing FLAs in order to sustainably maximize the potential yield of these leased areas. Conservatively if strategically implemented, this option may yield a net income of US$657 million in 10 years.

4.3 For ponds with expired FLAs and the pond existence is necessary based on bioeconomic analysis, reapply FLA and optimize fish yield (Samson and Rollon, 2008)

For brackish-water ponds with expired FLAs, the management strategy will require a balance between the importance of the pond area for fish production and the importance of these areas in restoring its natural ecological health and function (Janssen and Padilla, 1999; Barbier, 2000). The importance of the pond for fish production should not only be assessed in terms of what the operators can gain from it but also if its needed in a broader context of fish production in an area. For example, if an area is a major source of fisheries products and the non-operation of some ponds will cause a disruption on the supply, then production must be sustainably optimized to meet the target volume. This kind of assessment must be included in the procedures being followed by BFAR before renewing the lease agreement of expired FLAs. A more rigorous evaluation of the production efficiency of the pond must be developed to ensure the optimal use of the leased area. Pond operators must be required to submit a regular report of their production, as well as their income during the period when the lease agreement is in effect. This report must first be reviewed by BFAR before granting renewal.

4.4 For ponds with expired FLAs and production is suboptimal and not necessary, revert to mangrove areas thru natural revegetation or assisted planting (Samson and Rollon, 2008)

As mentioned earlier, under productivity of the pond may be brought about by the inappropriateness of the site for fish production (Stevenson et al, 1999). Problems such as the lack of supply of water and sedimentation may cause fish production to go down to unsustainable level. If this is the case, the site can be properly assessed for reversion to its original habitat. Natural revegetation will require much less labor and financial output as compared to assisted planting, however, there are cases when the site's modification will not anymore allow the recruitment and settlement of mangrove propagules in the area. As of 2007, there are around 39t hectares of brackish-water ponds with expired FLAs. Revegetating these ponds will greatly increase the percentage of rehabilitation efforts in the Philippines. Provinces with 85 to 100% of expired FLAs are Antique, Maguindanao, Lanao del Norte, Palawan, Basilan, Sulu, Davao Oriental, Sultan Kudarat, Northern Samar and Camarines Sur.

4.5 For active titled ponds, apply aqua-silviculture

Primavera in 2000 estimated that there are around 230t hectares of mangroves which had been converted to brackish-water ponds, using this number and subtracting the areas listed with FLA (59, 293), it will give us an estimate of around 74% (170,707 hectares) brackish-water ponds which are titled or undocumented. For these titled ponds, the management strategies will not be straightforward as the utilization of titled ponds rely greatly on its owners. The most rational option for these active title ponds is to sustainably operate the ponds to maximize the production potential of the area. Incentive mechanisms may be institutionalized to encourage titled pond owners to apply the two options most especially if these ponds are idle. These may be in the form of tax incentives and awards for sustainable operation, technical assistance to optimize production, and provision of seedlings for revegetation and others. One of the strategies that may benefit both the pond operators and government will be aqua-silviculture (Melana et al 2000a & b). Aquasilviculture promotes the mix of sustainable pond operation and the revegetation of some parts of the brackish-water ponds (Melana et al 2000a & b). Although this culture practice has been and are being practiced in some parts of the country (i.e. Aklan; Quezon), its ecological and economic benefits are not yet fully realized.

4.6 For idle titled ponds, fish yield may be optimized but pond size as in option 2 may be reduced to as small as possible and revert unutilized areas to mangrove forest

The option of revegetating idle ponds will greatly benefit pond operators and coastal communities as this may bring back mangrove goods and services and may potentially be a supplemental source of livelihood. One of the important mangrove services that the pond operators may consider is coastal protection. In the light of the looming impacts of climate change, mangroves will play a pivotal role in mitigating tsunami, strong waves and coastal erosion (Alongi, 2002, 2008; Dahdouh-Geubas et al, 2005; Gilman, 2006, 2008; Mc Leod & Salm, 2006; UNEP-WCMC, 2006).

Proper accounting and management of all brackish-water ponds in the country should be a top priority. A comprehensive and proper accounting of these titled ponds may help the

Philippine government formulate policies that will encourage environmentally sustainable fish production and resolve the fisheries and forestry utilization conflict in our mangrove areas.

Using available literature on the direct and indirect uses of mangroves (White & Cruz-Trinidad, 1998; Samonte-Tan & Armedilla, 2004; Walton, 2006), the TEV of the different options were estimated. Of the six management options that were proposed, the option that involves the reforestation of idle and unproductive ponds and the practice of aquasilviculture may bring about the highest total economic value at US$ 4.28 billion (at US$1 = Php 42.61) in 10 years. These two options provide opportunities to sustainably maximize the aquaculture potential and the variety of mangroves goods and services that this ecosystem could offer. In the aquasilviculture option, the economic goods will not only come from fish production from the pond itself but also to the recruitment and settlement of mangrove associated fish and organisms. The ecotourism potential of the system and its coastal protection value may also be restored thereby increasing the total economic value of the system.

5. Topographic/ hydrological conditions of these 'excess' areas in selected areas of the Philippines

What may pose a problem in the natural revegetation of the target pond site is its mean sea level elevation relative to the mean sea level elevation of the adjacent mangrove area (**Figure 8**). A study site in Lian, Batangas revealed that the pond site is 1 meter lower than that of the adjacent mangrove area. However this may not pose a serious problem for natural revegetation. Perhaps a natural in-filling by erosion should be allowed for some time, and elevation monitored. The first necessary step in the revegetation of the pond will be the opening of the tidal gates to allow for natural in-filling of sediment. This would also allow the recruitment and settlement of propagules however during the first months of revegetation, seedlings may not grow as they may either drown or get buried by inflowing

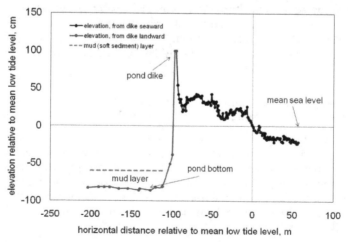

Fig. 8. Elevation map of the mangrove and pond site relative to mean low tide level at the study site in Lian, Batangas, Philippines.

of sediments (Krauss, 2008). As had been observed in other areas where tidal gates were destroyed, the pond elevation after some time levels with the adjacent mangrove forest, on the average this may take 1 to 2 years. This could then be followed by an observation period of where the seedlings will settle to determine if proper spacing of the seedlings to favor growth will be achieved. Natural recruitment of propagules in the pond site may not be a problem due to the abundance of seedlings near tidal gates as was observed in the study site in Lian, Batangas, (**Figure 9**). Assisted planting may be considered by replanting the propagules in the areas of the pond where elevation is relatively higher. Lewis (2005, 2009) in collaboration with various authors (1997, 2000) discussed in details the necessary steps for successful management and rehabilitation of abandoned or disused brackish-water ponds.

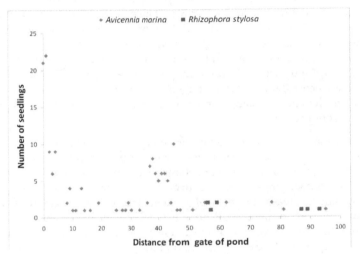

Fig. 9. Abundance of seedlings of *Avicennia marina* and *Rhizophora stylosa* from the seafront area of a mangrove forest to the gate of a brackish-water pond in Lian, Batangas, Philippines.

6. Conclusions

The inverse link between the remaining mangrove areas and existing brackish-water ponds in the Philippines had been well recognized for three decades now. The degraded state of mangrove areas is caused by its continued overexploitation for wood products, unregulated conversion for coastal development, inappropriate planting programs, and more importantly the absence or lack of more efficient and adaptive rehabilitation and conservation strategies. The proliferation of idle and underutilized brackish-water ponds is also an effect of the absence of efficient monitoring and evaluation system for the condition and status of these ponds. Laws and administrative orders for the rehabilitation and protection of mangroves areas as well as the responsible utilization of brackish-water ponds in the Philippines are in place, however, the implementing rules and regulations for these policies seemed to be lacking in coverage and effectivity. At present there are approximately 232,100 ha of brackish-water ponds, however only 59,923 ha have lease agreements from the Bureau of Fisheries and Aquatic Resources. At present, sixty-six percent of these FLAs are already expired. The remaining mangrove areas which were reportedly converted to brackish-water ponds are either titled or unaccounted. The appalling state of mangroves and brackish-water ponds in the Philippines results to continued loss of goods and services that these two systems could provide.

The strategies being proposed here aims to rationalize the environmental management of mangroves and brackish-water ponds in the Philippines. These strategies may be divided into three: 1) for brackish-water ponds with valid legal instruments and fish yield is still optimal, a more sustainable way of production must be adopted; 2) for ponds without legal instruments but fish yield is still bio-economically important, production must be sustainably optimized and legalized, and pond size may be reduced to as small as possible to maintain mangrove ecological health; and 3) for ponds with or without legal instrument and production is not anymore necessary, revegetation of the pond is being put forward. Of these options, the reforestation of idle and unproductive ponds and the practice of aquasilviculture may bring about the highest total economic values for these brackish-water ponds. The implementation of these options may not be straightforward and may need a conscious and concerted effort of different government agencies, academe and private institutions for an ecosystem based management approaches. These approaches should include mapping, inventory, status assessment and reprogramming and financing of existing management program.

With its current status, the vulnerability of mangroves to unsustainable anthropogenic activities (i.e. conversion to aquaculture areas, wood cutting, clearing for coastal developments) and the impacts of climate change may continue to degrade this ecosystem. Addressing these impacts may need the urgent rehabilitation of idle and unproductive ponds as these may decrease the level of vulnerability of these mangrove areas and the coastal communities behind them to the ancillary impacts of climate change. In all these, the government should take a proactive role in consolidating and monitoring these efforts to increase its efficiency and effectivity at the national scale as the present issues and problems on mangroves and brackish-water ponds in the Philippines cannot be appropriately addressed at the local scale.

7. References

Alongi, D.M. (2002). Present state and future of world's mangrove forests. *Environmental Conservation* 29(3): 331-349.

Alongi, D.M. (2008). Mangrove forests: Resilience, protection from tsunamis, and responses to global climate change. *Estuarine, Coastal and Shelf Science* 76, 1-13.

Barbier , E.B. (2000). Valuing the environment as input: review of applications to mangrove fishery linkages. *Ecological Economics* 35: 47-61.

Bureau of Fisheries and Aquatc Resources. (n.d.) List of FLAs duly issued by DA. In: *Bureau of Fisheries and Aquatic Resources,* June 27, 2011, Available from:
http://www.bfar.da.gov.ph/services/CRS_regulatory_svcs/listingoffla.htm

Cruz, P.S. (1997). Aquaculture Feed and Fertilizer Resource Atlas of the Philippines. *FAO Fisheries Technical Paper-T366,* 259pp.

Dahdouh-Guebas, F. et al. (2005). How effective were mangroves as a defense against the recent tsunami? *Current Biology* 15 (12): 443-447.

Department of Environment and Natural Resources – National Mapping and Resource Information Authority (DENR-NAMRIA). (2007). Regional Mangrove Statistics, unpublished.

Duke, N.C., Ball, M.C. & Ellison, J.C. (1998). Factors influencing biodiversity and distributional gradients in mangroves. *Global Ecology and Biogeography Letters* 7(1): 27-47.

Food and Agriculture Organization (FAO). (2003). Status and trends in mangrove area extent worldwide. By Wilkie, M.L. and Fortuna, S. Forest Resources Assessment Working Paper No. 63. Forest Resources Division. FAO, Rome. (Unpublished)

FAO. (2007). The world's mangroves, 1980-2005. FAO For. Pap. 153, 77 p.

Field, C.D. (1998). Rehabilitation of mangrove ecosystems: an overview. *Marine Pollution Bulletin* 37(8): 383-392.

Gilman, E., et al. (2006). Pacific Island Mangroves in a Changing Climate and Rising Sea. UNEP Regional Seas Reports and Studies No. 179. United Nations Environment Programme, Regional Seas Programme, Nairobi, KENYA.

Gilman, E. L. (2008). Threats to mangroves from climate change and adaptation options. *Aquatic Botany* (2008), doi:10.1016/j.aquabot.2007.12.009

Janssen, R., & Padilla, J.E. (1999). Preservation or Conversion? Valuation and evaluation of a mangrove forest in the Philippines. *Environmental and Resource Economics* 14: 297-331.

Krauss, K.W., et al. (2008). Environmental drivers in mangrove establishment and and early development: a review. *Aquatic Botany*, doi: 10.1016/j.aquabot.2007.12.014.

Lewis, R.R. III, Erftemeijer, P.L.A., & Sayaka, A. (2000). Mangrove rehabilitation after shrimp aquaculture: A case study in progress at the Don Sak National Forest Reserve. Southern Thailand. Thematic Review of Coastal Wetlands and Aquaculture, World Bank and NACA, Bangkok, Thailand.

Lewis, R. R., & Marshall, M. J. (1997). Principles of successful rehabilitation of shrimp aquaculture ponds back to mangrove forests. Programa/resumes de Marcuba '97, September 15/20, Palacio de Convenciones de La Habana, Cuba, 126pp.

Lewis, Roy R. (2005). Ecological engineering for successful management and rehabilitation of mangrove forests. *Ecological Engineering* 24: 403 – 418.

Lewis, R.R. (2009). Methods and criteria for successful mangrove forest rehabilitation. In: Perillo, G.E., Wolanski, E., Cahoon, D.R., Brinson, M.M. (Eds), Coastal Wetlands: *An Integrated ecosystem approach*. Elsevier, pp 787-800.

McLeod,E., & Salm, R.V. (2006). Managing Mangroves for Resilience to Climate Change. IUCN, Gland, Switzerland. 64pp.

Melana, D.M., Melana, E.E., & Mapalo, A.M. (2000a). Mangrove Management and Development in the Philippines. Oral presentation at *Mangrove and aquaculture management*, 14 – 16 February, 2000, Kasetsart University Campus, Bangkok, Thailand.

Melana, D.M., et al. (2000b). Mangrove Management Handbook. Department of Environment and Natural Resources, Manila, Philippines through the Coastal Resource Management Project, Cebu, Philippines, 96 p.

Polidoro, B.A. et al. (2010). The Loss of Species: Mangrove Extinction Risk and Geographic Areas of Global Concern. *PLoS ONE* 5(4): e10095. doi:10.1371/journal.pone.0010095.

Primavera, J.H. (1991). Intensive prawn farming in the Philippines: ecological, social, and economic implications. *Ambio* 20 (1): 28-33.

Primavera, J.H. (1995). Mangroves and brackish-water pond culture in the Philippines. *Hydrobiologia* 295: 303-309.

Primavera, J. H. (2000). Development and conservation of Philippine mangroves: institutional issues. *Ecological Economics* 35 (Special Issue): 91-106.

Primavera, J.H. (2006). Overcoming the impacts of aquaculture on the coastal zone. *Ocean and Coastal Management* 49: 531-545.

Primavera, J. H. & Agbayani, R.F. (1997). Comparative strategies in community-based mangrove rehabilitation programmes in the Philippines. In: Hong, P.N., Ishwaran, N., San, H.T., Tri, N.H., Tuan, M.S. (Eds.), Proceedings of Ecotone V, Community Participation in Conservation, Sustainable Use and Rehabilitation of Mangroves in Southeast Asia. UNESCO, Japanese Man and the Biosphere National Committee and Mangrove Ecosystem Research Centre, Vietnam, pp 29-243.

Primavera, J.H., & Esteban, J.M.A. (2008). A review of mangrove rehabilitation in the Philippines: successes, failures and future prospects. *Wetlands Ecology and Management* 16 (3): 173-253.

Primavera, J.H., Rollon, R.N., & Samson, M.S. (2012). The pressing challenges of mangrove rehabilitation: pond reversion and coastal protection. *Treatise on Estuarine and Coastal Science*.

Primavera, J.H. et al. (2004). Handbook of mangroves in the Philippines - Panay. SEAFDEC Aquaculture Department (Philippines) and UNESCO Man and the Biosphere ASPACO Project, 106 pp.

Samonte - Tan, G. & Armedilla, M. C. (2004). Sustaining Philippine Reefs: National Coral Reef Review Series No. 2 - Economic Valuation of Philippine Coral Reefs in the South China Sea Biogeographic Region. UNEP - GEF South China Sea Project, Marine Science Institute, University of the Philippines, Quezon City, Philippines, vi + 38pp.

Samson, M.S., & Rollon, R.N. (2008). Growth performance of planted mangroves in the Philippines: revisiting forest management strategies. *Ambio* 37(4): 234-240.

Spalding, M., Kainuma, M., & Collins, L. (2010). World Atlas of Mangroves. Earthscan Publication, U.K., 319pp.

Stevenson, N.J., Lewis, R.R. III, & Burbridge, P.R. (1999). Disused shrimp ponds and mangrove rehabilitation. In: Streever, W. (Ed.), An International Perspective on Wetland Rehabilitation. Kluwer, Netherlands, pp. 277-297.

UNEP-WCMC. (2006). In the front line: shoreline protection and other ecosystem services from mangroves and coral reefs. United Nations Environment Programme - World Conservation and Monitoring Centre, Cambridge, United Kingdom, 33 pp.

Vermaat, J.E., & Thampanya, U. (2006). Mangroves mitigate tsunami damage: A further response. *Estuarine, Coastal and Shelf Science* 69: 1-3.

Walton, E.M., et al. (2006). Are Mangroves worth replanting? The direct economic benefits of a community-based reforestation project. *Environmental Conservation* 33(4): 335-343.

White, A. T., & Cruz - Trinidad, A. (1998). The Values of Philippine Coastal Resources: Why Protection and Management are Critical. Coastal Resource Management Project, Cebu City, Philippines, 96pp.

Yap, W.G. (2007). Assessment of FLA holdings in four pilot regions. Strategy for Sustainable Aquaculture Development for Poverty Reduction, Philippines (ADTA 4708-PHI), PRIMEX, Manila. Unpub. report. 33 pp.

4

Manila Clam (*Tapes philippinarum* Adams & Reeve, 1852) in the Lagoon of Marano and Grado (Northern Adriatic Sea, Italy): Socio-Economic and Environmental Pathway of a Shell Farm

Barbara Sladonja[1], Nicola Bettoso[2], Aurelio Zentilin[3],
Francesco Tamberlich[2] and Alessandro Acquavita[2]
[1]*Institute of Agriculture and Tourism Poreč, Poreč,*
[2]*ARPA FVG-Osservatorio Alto Adriatico, Trieste*
[3]*Almar Soc. Coop. Agricola a.r.l., I-33050 Marano Lagunare (UD)*
[1]*Croatia*
[2,3]*Italy*

1. Introduction

Manila clam is a subtropical to low boreal species of the western Pacific, distributed in temperate areas of Europe. The natural populations are distributed in the Philippines, the South China and China Seas, Yellow Sea, Sea of Japan, the Sea of Okhotsk, and around the Southern Kuril Island (Scarlato, 1981). Its culture was initiated in those areas from the initial traditional fishing activities by the collection of wild seeds.

As a species of commercial value, Manila clam has been introduced to several parts of the world, to become permanently established in several areas. The species was accidentally introduced the 1930's to the Pacific coast of North America along Pacific oyster *Crassostrea gigas* seed import (Chew, 1989). The species naturally spread the Pacific coast from California to British Colombia (Magoon & Vining, 1981). Besides public fisheries, hatchery production has facilitated Japanese carpet shell culture along the Pacific coastline. Manila clam was also transferred from Japan to Hawaiian waters early in the 20th century, where wild populations still occur. Overfishing and irregular yields of the native (European) grooved carpet shell, *Ruditapes decussatus*, led to imports of *R. philippinarum* into European waters.

In 1972, the species was introduced into France by a commercial hatchery where they cultivated since the early 1980's (Goulletquer, 1997). The aquaculture development was facilitated by commercial hatcheries and additional imports from the United Kingdom using broodstock from Oregon (USA), resulted with numerous transfers within the European Union borders (Portugal, Italy, Ireland, Spain). Moreover, aquaculture experiments resulted in seed imports into Belgium, Germany, Israel, Tahiti, Tunisia, (Cesari & Pelizzato, 1985; Shpigel & Friedman, 1990).

In March 1983 the species was introduced in Italy in the Venice Lagoon (South Basin) by purchasing the seed from an English hatchery (Breber, 1985) (Fig. 1).

1986, Laguna di Marano e Grado (Zentilin, 1987)

1983, Laguna di Venezia (Breber, 1985)
1985, Sacca degli Scardovari (Milia, 1990)
1986, Sacca di Goro (Paesanti, 1990)

1989, Lago di Sabaudia
(Di Marco *et al.*, 1990)

1985, Sardegna
(Cottiglia *et al.*, 1988)

Fig. 1. Introduction of Manila clam (*Tapes philippinarum*) in Italy (Zentilin et al., 2007)

The promising results and the interest shown by many in this new zooculture, led to the spread of sowing in many transitional environments of Veneto (Lagoon of Venice, Lagoon of Caorle and the Po River Delta) (Breber, 1985; Giorgiutti et al., 1999), Emilia Romagna (Sacca di Goro and Sacca di Scardovari) (Paesanti, 1990; Milia, 1990), Friuli Venezia Giulia (Lagoon of Marano and Grado) (Zentilin, 1987), Sardinia (Cottiglia et al.,1988), and Lazio (Lake Sabaudia) (Di Marco et al., 1990). Following the large aquaculture hatchery based on developments in Europe over the 1980s, natural reproduction resulted in a geographical expansion of wild populations, particularly in Italy, France, and Ireland, where Japanese carpet shells have proved to be hardier and faster growing than the endemic *R. decussatus*. Consequently, *R. philippinarum* populations are now the major contributor to clam landings in Europe, and are the focus of intensive public fisheries, competing with aquaculture products in several rearing areas.

In these areas the species has acclimatized, reproduced and distributed at all sites, more environmentally favorable. The colonization of new environments has been fast and that makes the Manila clam the most economically important species.

The Manila clam (*Tapes philippinarum*) culture was preferred to that of the indigenous, *Tapes decussates* due to high growth rate, less difficult in obtaining seed from controlled

Manila Clam (Tapes philippinarum Adams & Reeve, 1852) in the Lagoon of Marano and Grado
(Northern Adriatic Sea, Italy): Socio-Economic and Environmental Pathway of a Shell Farm

53

reproduction and for better tolerance to variations of temperature, salinity and quality substrate that it presents with respect to local species.

2. Biological features and life cycle of Manila clam

Manila clam *(Tapes philippinarum* or *Venerupis philippinarum)* is an edible species of saltwater clam, a marine bivalve mollusk in the family *Veneridae*, the Venus clams. The common names of the species include "Manila clam", "Japanese littleneck", "steamer clam", "Filipino Venus", "Japanese cockle", and "Japanese carpet shell". Its shell is solid, equivalve, inequilateral, beaks in the anterior half, somewhat broadly oval in outline. Ligament inset, not concealed, a thick brown elliptical arched body extending almost half-way back to the posterior margin. Lunule elongate heart-shaped, clear though not particularly well defined, with light and dark brown fine radiating ridges. Escutcheon reduced to a mere border of the posterior region of the ligament. Sculpture of radiating ribs and concentric grooves, the latter becoming particularly sharp over the anterior and posterior parts of the shell, making the surface pronounced decussate. It has three cardinal teeth in each valve; centre tooth in left valve and centre and posterior in right, bifid. No lateral teeth. Pallial sinus relatively deep though not extending beyond the centre of the shell; it leaves a wedge-shaped space between its lower limb and the pallial line. Margins are smooth. Extremely variable in color and pattern, white, yellow or light brown, sometimes with rays, steaks, blotches or zig-zags of a darker brown, slightly polished; inside of shell polished white with an orange tint, sometimes with purple over a wide area below the umbones. The features most diagnostic for the identification of this species are the following: the inner ventral margin of the shell is smooth; the ligament is prominent and elevated above the dorsal margin. In the living animal, the siphons are separated at the tips. Water is drawn in and out of a clam by siphons that protrude from the posterior end of the shell. In this species, the siphons are mostly fused, and are only separate at the tips. The siphons are short relative to some other clams, which means that the clam lives burrowed only a shallow distance under the surface of the substrate. Short siphons are what inspire the common name "littleneck clam" (Bourne, 1982).

T. philippinarum is a dioecious animal (Bardach et al., 1972; Chew, 1989; Eversole, 1989; Devauchelle, 1990). In natural populations *T. philippinarum* becomes sexually mature in the first to the third year of age. *T. philippinarum* are strictly gonochoric and their gonads are represented by a diffused tissue closely linked to the digestive system. The period of reproduction varies, according to the geographical area; spawning usually occurs between 20-25 °C. A period of sexual rest is observed from late autumn to early winter. Gametogenesis in the wild lasts 2-5 months, followed by the spawning. A second spawning event may occur in the same season, 2-3 months later. The pre-winter recovery phase facilitates energy build up, by filtering seawater still rich in organic matter and phytoplankton. Temperature and feeding are the two main parameters affecting gametogenesis, which can be initiated at 8-10 °C and is accelerated by rising seawater temperature. Its duration decreases from 5 to 2 months between 14 and 24 °C. Within this temperature range, *T. philippinarum* are ready to spawn. Although the optimal temperature is between 20 and 22 °C, 12 °C is the minimum threshold below which this species cannot spawn efficiently. Food availability influences the amount of gametes produced. Larval development lasts 2 to 4 weeks before spatfall. Settlement size is between 190 and 235 µm in shell length. Many external factors regulate spatfall success in the wild, such as temperature, salinity and currents. Larval movement mainly depends on wind driven and tidal currents. Adding pea gravel and small rocks can facilitate species recruitment in natural setting areas. The larvae settle by attaching a byssus to a pebble or piece of shell. Regarding the *Tapes* longevity, Ponurovski (2008) has reported that the maximum recorded age of *T. philippinarum* was 7 years, but the maximum longevity of this species that has ever been reported for Melkovodnaya Bay (Russia) is 25 years.

The reproductive cycle in the northern Adriatic lagoons is quite similar (Da Ros et al., 2005): the period of sexual rest lasts from October to January, when first specimens start the gametogenesis. Sexual maturity is observed from April and the first spawning occurs already in May, the last in September. The theoretical growth curves depend on the period of larval settlement. When it occurs during the spring, Manila clam can reach about 47 mm in length after two years (Pellizzato et al., 2005). The maximum length recorded was 78 mm in Sacca di Goro (Po River Delta), whereas in the Lagoon of Marano the record was 71 mm, 136 g in total weight and about 7 years old.

3. Statistics

3.1 Main producer countries and production statistics

Cultivation of Manila clam, with the world production of 3 million tons/year makes it economically the most important shell species. In fact, its production represents the 20% of global world shell market (FAO, 2011).

At the global level, since 1991, global Japanese carpet shell production has shown a huge expansion, by a factor of nearly six times. China is by far the leading producer (97.4 % in 2002). Disease factors have impacted production in some other countries, notably in the Republic of Korea. In the decade 1993–2002, production in that country varied between 10,000 and 19,000 tones.

Production in Italy, following its introduction, the development of wild populations, and the consequential increase in seed supply, is the second highest in the world, followed by United States of America, Spain and France. Extensive production also occurs in Japan but is not reported within this specific statistical category, being included within clams (FAO, 2011).

In Europe, 90% (50,000 tons /year) of production derives from Italy, 6-8% from Spain (4,000 tons/year) and 2% from France (1,000 tons/year) (Turolla, 2008).

Italian clam production is mostly concentrated in the lagoons of the North Adriatic. Northern Adriatic production contributes to overall Italian production by up to 95% (Orel et al., 2000; Zentilin et al., 2008). This species is one of the most popular and profitable molluscs of lagoon and coastal sites in the Mediterranean. Production areas in the northern Adriatic (Lagoon of Venice, Lagoon of Marano and Grado, River Po mouth in the Region Emilia Romagna, River Po mouth in the Region of Veneto) reached its peak in 1998 and 1999 with over 60,000 tons. In the Lagoon of Marano and Grado the highest production was registered in 1996 with over 1,500 tons. In 2010 the production in the Lagoon of Marano was 1,042 tons, of which 74% derives from aquaculture in about 130 hectares of lagoon. Approximately 20 persons are employed in the two main shell farm business operators: ALMAR Society founded in 1995 and Molluschicoltura Maranese Society (begin of '90th), which took over the shell farm plants of the previous Aquamar Society, founded in 1988.

Manila Clam (Tapes philippinarum Adams & Reeve, 1852) in the Lagoon of Marano and Grado
(Northern Adriatic Sea, Italy): Socio-Economic and Environmental Pathway of a Shell Farm

55

3.2 Market and trade

About 70% of Italian production is consumed internally, while the remaining is exported, mostly towards European countries, Spain primarily. Due to increased consumption during summer months, in just two months (August and December), 20-25% of all Italian production is collected and commercialized (Turolla, 2008). Consumption is almost exclusively directed to live product trade, only in the last years product processing is starting to take part on the market.

4. Lagoon of Marano and Grado

The Marano and Grado Lagoon is located in the northern Adriatic Sea covering an area of about 160 km² with a length of nearly 32 km and an average width of 5 km (Fig. 2).

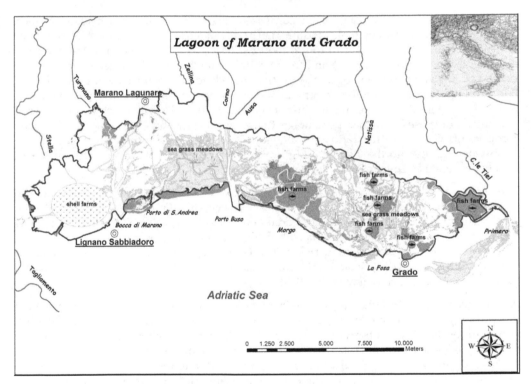

Fig. 2. The Lagoon of Marano and Grado

It is the more eastern among all the delta and lagoon systems on the Adriatic coast (Po River Delta, Lagoon of Venice and Lagoon of Caorle). This wetland system is delimited, moving westward, by the Isonzo and Tagliamento Rivers deltas, and is separated from the Adriatic Sea by the presence of six barrier islands (1.6 to 6 km long, for a total length of 20 km; 1-2 km wide). These latter roughly subdivide the lagoon in hydraulic sub-basins (Marano, S. Andrea, Porto Buso, Morgo, Grado and Primero; ARPA-FVG, 2006).

The whole catchments area of the lagoon covers about 1880 km² with 65 municipalities and a rivers network of 743 km. Rivers flow into the lagoon mainly in the western sector (Marano) draining waters coming from the spring line. The estimated overall amount of average freshwater discharge is about 70-84 m³ s⁻¹ (ARPA-FVG, 2006; Ret, 2006) and is mostly due to the Stella River, particularly important for its peculiar estuarine habitats, the Cormor River and the Corno River, that receives the water of the Aussa River before entering the lagoon. Salinity is strongly influenced in the areas close to the river mouths where very low values (from 2 to 7) are often recorded. Conversely, a sudden increase is observed moving towards the tidal inlets, where values between 24 and 36 (typically marine) were recorded (Brambati, 2001). Temperature gradients and other biogeochemical parameters (i.e., dissolved oxygen, nutrients, chlorophyll) are characterized by high spatial and temporal variability (Acquavita et al., 2010a), with some extreme phenomena such as ipoxic/anoxic conditions recorded in the more confined areas especially in case of a high organic carbon content coupled to a scarce water renewal.

The sedimentary input of small rivers consists mostly of fine suspended material washed out from the groundwater table (Brambati, 1972), whereas the most important source of sediments comes from the sea, through the tidal inlets, as a contribution of the two river deltas (silty and clayey particles) and from erosion of the barrier islands (sands) (Brambati, 1970). Grain-size distribution observed for the whole lagoon is characterized by medium to fine sands prevailing along the both main and secondary channels. The fine pelitic fraction is widely dispersed in the basin and varies from 52% to 100%. The silty fraction (62.5-3.9 µm) normally comprises a higher percentage than the clay fraction (<3.9 µm) in the pelitic component. Generally, the ratio fine-silt/clay (2-16/<2 µm) identifies areas affected by freshwater inputs into the lagoon: a high ratio indicates an area of fresh water input and vice versa (Piani & Covelli, 2000; Piani et al., 2005).

The main sub-basins identified in the past are characterized by several morphologies. The presence of barriers, sandbanks, saltmarshes, channels, intertidal/subtidal flats and tidal creeks are strictly dependent from both tidal and hydraulic regimes. The Marano Lagoon is a semi-enclosed shallow tidal basin, with a few marshes and several channels, receiving freshwaters from several adjacent rivers. The Grado Lagoon is shallower, has a series of morphological relieves (islands) and marshes, and receives freshwater from a single tributary, the Natissa river (Marocco, 1995).

As already reported for other lagoon environments this system is clearly dynamic, being the morphology subjected to relevant changes derived from both natural and human influences. Fontolan et al. (2009) observed a remarkable decrease of saltmarshes surface, which occurred especially in the last decades (1954-1990), in the central-western areas (i.e. S. Andrea, Grado and Porto Buso sub-basins), probably due to relative sea level rise, erosive processes and land reclamation. Actually, the main salt marshes are located between Marano and S. Andrea sub-basins and in Morgo. Fringing marshes are also located in the inner parts of the Marano sub-basin.

However, the influence of human activities on lagoon morphology cannot be neglected. In order to protect the main inlets breakwaters were constructed, along with the creation of the Porto Nogaro harbor, which is connected by means of a lagoon channel (Aussa Mare), up to 7.5 m of depth, with the open sea for shipping purposes. In addition, stabilization of the barrier islands and several docks for marinas were realized.

Manila Clam (Tapes philippinarum Adams & Reeve, 1852) in the Lagoon of Marano and Grado
(Northern Adriatic Sea, Italy): Socio-Economic and Environmental Pathway of a Shell Farm

57

Lagoon channels actually represent a 100 km network, partly navigable, connecting the lagoon to the northern Adriatic Sea, with waterways close to the main industrial and tourist areas.

4.1 Hydraulic regime

Hydraulic circulation and diffusive processes in lagoon environments exert a pivotal role in most of the physical and biogeochemical processes. Interannual and interseasonal variations of hydraulic forcing greatly contribute to biogeochemical processes through dispersion of nutrients and pollutants (Rigollet et al., 2004) and changing basic physico-chemical parameters (i.e., turbulence, dissolved oxygen, pH, salinity). During the first half of the last century some investigations concerning hydrodynamic of the lagoon were conducted; however, variation in both physical and chemical parameters were firstly described in order to define water circulation pattern by Dorigo only in 1965. Recently, more sophisticated 2D models were applied by Bosa & Petti (2004) and Petti & Bosa (2004) for clarifying the sediment transport and the dispersion of a dissolved pollutant in the lagoon environment. Ferrarin et al. (2010) employed a finite element modelling system (named SHYFEM, http://www.ve.ismar.cnr.it/shyfem) developed at ISMAR-CNR (Umgiesser et al., 2003; 2004) to describe the water circulation taking into account different forces such as tides, wind, rivers and sea-lagoon exchanges. One of the main outputs of this work was the comparison of the simulation against water levels, salinity and temperature data collected in several stations inside the lagoons. The lagoon has a strong tidal influence with several basins separated by the watersheds where exchange of water between neighboring basins occurs (Umgiesser, 1997). In a simulation proposed following Solidoro et al. (2004) employing different tracers, the lagoon can be roughly divided in three subbasins. Taking into consideration areas influenced by riverine inputs, a clear west-east salinity gradient was found (from ~20 to~ 34 in western and eastern region, respectively) with a narrow band (1-2 km width) where a rapid change from mesohaline to euhaline conditions occur. The eastern part of the lagoon presents a physical division due to the bridge connecting the main land to the city of Grado.

4.2 Conflicts in coastal areas

The Marano and Grado Lagoon is one of the most important ecological systems of the North Adriatic Sea, assuring connectivity with some psammophile and halophile habitats of the coast. Nowadays, most of the natural areas are included in the Regional System of Protected Areas (S.A.R.A.), as Communitary Importance Sites (CIS) and Special Protection Zones (SPZ). The area includes some historical sites designed to protect wildlife migration submitted to the Ramsar Convention in 1971, and, following the implementation of the Habitats Directive (92/43/EEC transposed in Italy by DPR Sept. 8, 1997 No 357), concerning the protection of biodiversity, the entire basin has been identified in the survey sponsored by the state called "Natura 2000 "as a site to be included among the sites of Community importance (SCIs - IT3320037). Moreover, the lagoon is an economical important source for the inhabitants.

Fishing involves actually 245 fishermen and 200 persons in charge of marketing in Marano Lagunare, whereas in Grado tourism represents the leading sector (over 1,700 persons), and is almost exclusively carried out from April to October. Aquaculture (lagoon fish farming) is a common historical practice for the upper Adriatic. Within the lagoon system 17 fish rearing ponds (320 hectares in Marano) and 38 (1,088 hectares in Grado) are still active. Due to the shallow depth of the basins, arise of temperature and salinity linked to hypoxia may occur during summer periods, whereas occasionally frosts occur during winter.

On the other hand, the lagoon has been declared one of the polluted sites of national interest (SIN; Ministerial Decree 468/01) because of its high level of sanitary and environmental risk due to several factors. In this context one of the main concerns is represented by the high concentration of mercury (Hg) in sediments and its related neurotoxicity whenever it is in the food chain. SIN covers about 13,500 ha partially overlaps the SCI, and is actually subject to environmental characterization, emergency safety measures, restoration, rehabilitation plans and monitoring activity.

The Environmental Protection Agency of Friuli Venezia Giulia (ARPA FVG) is involved in a monitoring network according to Directive 2000/60/CE (European Union, 2000), which is applied in Italy following the Legislative Decree 152/2006. The WFD aiming to monitor and solve the problems connected with quality of the water, to protect and to enhance the restore of all the water bodies (inland surface waters, transitional waters, coastal waters and ground waters in Europe). The final goal is to achieve a "good ecological and chemical status" by the year 2015.

The data obtained represent an implementation of the past investigations (since the end of the 80s) conducted by the Region FVG in a monitoring system of superficial water, where physical, chemical and bacteriological data were recorded. GIS based analysis of data collected from about 20 monitoring stations for the period 2000 – 2005 show that the lagoons is characterized by a good diurnal oxygenation. On the other hand, a significant nutrient supply pose a risk of distrophy close to river mouths with the induction of algal blooms and possible oxygen falls during the night. Moreover, these areas are subject to high nitrogen concentrations, which go beyond the limits stated into the Nitrate Directive.

4.2.1 Agriculture

Agriculture and breeding are activities connected to the quality of the Marano and Grado Lagoon mainly through the related drainage basin of cultivated land. Rural economy of Friuli Venezia Giulia is almost placed in the Lower Friulan Plan, where, during the last century, 50,000 hectares of wetlands have been reclamed into cultivated areas. Nowadays, 70% of the agricultural land is arable and mostly planted with soybean, Indian corn, beetroot and cereals (wheat and barley). There are also several vineyards or specialized fruit farming.

The main risks posed by agricultural activities are pesticides and fungicides releases into the lagoon system. The irrigation of the plantations, especially during the summer periods, implies a serious water withdrawal, and the lack of vegetation enhances the problem of erosion. In the considered territory, there are several breeding farming. The wastes of this activity are used as fertiliser and, as a consequence, part is carried into the lagoon (Friuli Venezia Giulia Region, 3th Interim Report "S.A.R.A. Regional System of Protected Areas").

The main river courses of the area are loaded with the discharge of the channels belonging to the agricultural drainage system. Thus, it results in a close relationship between agriculture activities and the quality of waterways and lagoon ecosystem. The release of

Manila Clam (Tapes philippinarum Adams & Reeve, 1852) in the Lagoon of Marano and Grado
(Northern Adriatic Sea, Italy): Socio-Economic and Environmental Pathway of a Shell Farm

59

nutrients from agriculture are closely linked to hydrological processes as they are conveyed during rainfall in the catchment and load significant amounts of nitrogen and phosphorus supplied to the soil with fertilizers. Due to their patterns of adsorption and release from the soil nitrogen is mainly present as $N\text{-}NO_3^-$ ion when reach the lagoon system, whereas P is tied to the particle in suspension.

4.2.2 Industry

Besides its environmental and natural significance, the lagoon hosts some important socio-economic activities. Two industrial sites are located in San Giorgio di Nogaro and Torviscosa, whereas a commercial harbour sited in "Porto Nogaro" is connected to the lagoon by mean of channel along the Ausa-Corno River. The industrial plants of the Aussa-Corno (ZIAC) industrial includes 87 factories and about 2500 persons are actually employed. Two separate districts are placed in the whole area. The largest area (498 ha) extends in San Giorgio di Nogaro municipality along 6,5 km of the Aussa river up to Marano and Grado lagoon. An area lies at the confluence between Aussa and Corno rivers; a part is located near the Ausa river. A second area is in Torviscosa municipality and is occupied by "Caffaro" industries (130 ha). There are 320 ha not yet occupied. The "Porto Nogaro" serves the industrial area and moves about 1,500,000 tons/year. Categories, employees and area of the factories in the ZIAC are reported in the following table (from: http://www.dest.uniud.it/dest/eventi/watergeography/lagoon_of_marano_tesis.pdf).

Category	N° employees	Area (m²)
Mechanical engineering	176	93,297
Iron and steel (production)	41	164,269
Iron and steel (manufacturing)	554	890,238
Chemical (production)	508	1,255,359
Chemical (articles)	555	674,056
Textile	66	45,340
Food	141	148,457
Naval/nautical	116	336,821
Wood	20	17,240
Stone processing	41	32,570
Total	2218	3,657,647

Table 1. Categories, employees and area of the factories in the ZIAC

The Torviscosa site still represents the main source of pollutants in the area. The consequent contamination regards both soils (manly the upper meter) and groundwater (Menchini et al., 2009). The main concern is mercury (Hg), due to the chlor-alkali plant that became operative in 1949 and that actually is not-operative. Moreover, the occurrence of other contaminants cannot be neglected. The presence of heavy metals (Cd, Zn, As, Pb, Cu), polychlorinated dibenzodioxins (PCDDs) and furans (PCDFs), hydrocarbons, polycyclic aromatic hydrocarbons (PAHs), is closely related to the location of the different chemical productions that took place in the industrial site. San Giorgio di Nogaro shows a less critical situation with soil contamination related to waste disposal and industrial processes, mainly heavy metals and to a lesser extent PAHs and PCBs. In a similar manner, the environmental quality of the Corno River seems better than the Aussa one, both in terms of contaminants and measured concentrations.

4.2.3 Exploitation of aquatic resources

The Lagoon of Marano and Grado hosts the two most important fishery ports of the Region Friuli Venezia Giulia: Marano Lagunare and Grado. Together they represent almost 80% of the fishery fleet of the region as number of vessels. Fishermen operate at sea by trawling, seine, hydraulic dredge for bivalves (mainly *Chamelea gallina*, *Callista chione* and *Ensis minor*), trammel and gill nets, longlines, fish traps adapted to catch cuttlefishes (*Sepia officinalis*) and mantis (*Squilla mantis*). Most of fishermen in Marano Lagunare work both at sea and lagoon; in the latter case they use prevalently a very old method to catch fish in the northern Adriatic lagoons: a barriers system ending in the fyke nets (locally named *cogolli*). This very ancient method is tightly linked to the tide regime. Moreover in the lagoon there is a so much ancient tradition (arising to 16th century in the Lagoon of Grado) to farm fishes inside pools such as sea bream (*Sparus aurata*), sea bass (*Dicentrarchus labrax*), eel (*Anguilla anguilla*) and mullet (*Mugil* sp.).

The spreading of Manila clam in the Adriatic lagoons led to a consistent abandonment of traditional fishery, because of the massive abundance of this species, its easiness of harvesting and overall the considerable market of this resource. In this way at the beginning of 90th almost 50% of fishermen in Marano Lagunare started to harvest these clams. In the first time they were harvested by hand from spring to autumn and a maximal daily quota per person of about 50 kg was established. In 1992 a special dredge was introduced to collect clams during the winter in the Lagoon of Marano. This gear was studied and built in Marano Lagunare and it is locally known as "maranese rake" (Fig. 3.). This method permitted to harvest Manila clams in the cold season and these gears were hauled on the shallow lagoon bottoms by boats with a power engine ranging between 80 and 120 Hp. A quota of 50 kg/day was established for each boat, excepting for about twenty days during the Christmas celebrations when a quota of 70 kg/day was permitted (Zentilin & Orel, 2009). No specific limitations were established to access the Manila clam natural banks with this method of fishery. Thus until December 2006 almost 150 fishermen could harvest freely this resource from October to March using this mechanical gear in the most part of the Lagoon of Marano. In the Lagoon of Grado Manila clam had not a massive spreading such as Marano basin and this species is manually harvested only by local or leisure fishermen; although an illegal harvesting and sale occur nowadays in both lagoons.

The Table 2. shows how the exploitation with "maranese rake" increased enormously since 1995, when the contribution of farm to the total production in the market of Marano Lagunare was only 3%, whereas the biggest production was recorded in 1996. In those years each natural bank of Manila clams was freely overexploited until the biomass dropped below 0.01 or even 0.005 kg/m^2. This impulsive and uncontrolled exploitation of the resource, accompanied to mobilization of sediments and potential contaminants, led the authority to stop definitively this free practice with the rake in January 2007, after many years of tacit assent. This sudden decision caused social disorders and resentments between the citizens of Marano Lagunare where, due to the general crisis of marine resources too, half of fishermen community was employed till then to this kind of harvesting. This occurs

Manila Clam (Tapes philippinarum Adams & Reeve, 1852) in the Lagoon of Marano and Grado
(Northern Adriatic Sea, Italy): Socio-Economic and Environmental Pathway of a Shell Farm

61

Fig. 3. The "maranese rake"

Year	Shell farm (tons)	Fishery (tons)	Total Manila clam production (tons)	%Shell farm
1988	3		3	100
1989	0.9		0.9	100
1990	9		9	100
1991	16		16	100
1992	35		35	100
1993	80	70	150	53
1994	70	150	220	32
1995	30	1,000	1,030	3
1996	60	1,560	1,620	4
1997	25	1,230	1,255	2
1998	140	800	940	15
1999	310	760	1,070	29
2000	911	463	1,374	66
2001	647	641	1,288	50
2002	535	468	1,003	53
2003	595	378	973	61
2004	417	602	1,019	41
2005	680	531	1,211	56
2006	368	300	668	55
2007	375	193	568	66
2008	204	169	373	55
2009	699	145	844	83
2010	772	270	1,042	74

Table 2. Production of Manila clam in tons in the Lagoon of Marano and percentage
contribute of shell farm to the total production

especially during the winter when Manila clam represents one of few resources accessible and capable to ensure a profit to the families.

In such a climate the pioneers of the cultivation of Manila clam had to operate for the development of their activity. From the beginning they embraced the principle on which the lagoon must be cultivated, to ensure a profitable activity for the fishermen community in the long time, thus minimizing the impact in the environment. Of course a lot of fishermen fell in contrast against farmers and now a minority of them sustains to be entitled for the free use of the "maranese rake" in their lagoon. In fact they adduce the principle on which only the local fishermen know how to manage their lagoon, on the basis of old knowledge inherited by fathers and concessions got in the ancient times for their exclusive utilization of the lagoon resources.

The challenge today is to promote a cultural revolution inside the Maranese community in order to gradually convert the clam fishermen toward the sustainable aquaculture, thus introducing the new conceptions of the shell farm business operator. In any case the free harvesting by hands is nowadays authorized, although most of the production derives from aquaculture. However the farmers must maintain continuously a vigilance staff to prevent the frequent illegal harvesting, inside the areas devoted to the Manila clam cultivation.

4.3 Heavy metals in the Lagoon environment

Natural processes such as leaching of the rocks, erosion and geothermal activity, and human activities such as fuel use fossil, mineral fusion, industrial and oil contribute to the diffusion of heavy metals in the atmosphere, water, soils, river and coastal sediments (Solomons & Föerstner, 1984). Sediments are known to be one of the most important site of deposition and a secondary source of trace elements to the marine environment (Leoni & Sartori, 1997).

The distribution of heavy metals in sediments of the Marano and Grado Lagoon was investigated since '70s (Stefanini, 1971), but only recently an evaluation of the anthropogenic influence on contamination levels was considered (Mattassi et al., 1991). The Authors pointed out the huge amount of Hg, with values very high (from 0.61 to 14.01 $\mu g\ g^{-1}$) if compared with other Mediterranean systems. On the other hand, Cd, Cr, Cu, Ni, Pb and Zn showed no significative enrichment and an homogenous distribution in the whole system. Due to this, Hg distribution, speciation and mobility is the main concern in the whole lagoon environment. The element can exert neurotoxic effects on humans and can be transformed in the more toxic and potentially bioaccumulable organometallic form monomethylmercury (MeHg). The huge amount of Hg found is mainly due to the particulate inputs from the Soca/Isonzo drainage basin. The River is responsible for the presence of Hg into the northern Adriatic Sea since the 16th century, due to its transport of cinnabar rich tailings from the Idrija mining district (North-West of Slovenia; Covelli et al., 2001; 2008), which Hg production was second only to Almaden (Spain). The Idrija Hg mine operated for nearly 500 years, until it definitively closed in 1996. Nevertheless, Hg is still being delivered through river flow to the Gulf of Trieste and the adjacent Marano and Grado Lagoon (Covelli et al., 2007). Within the lagoon Hg contents range from 2.34 to 10.6 $\mu g\ g^{-1}$ and progressively decrease westwards. Moreover, the Marano and Grado Lagoon experienced the input from the chlor-alkali plant (CAP) sited in Torviscosa. Here, Hg was employed for electrolytic production of Cl_2. The plant became active in 1949 and it was estimated that a total amount of 186,000 kg of Hg (with a maximum of about 20 kg day^{-1}) was deliberately discharged into the

Manila Clam (Tapes philippinarum Adams & Reeve, 1852) in the Lagoon of Marano and Grado
(Northern Adriatic Sea, Italy): Socio-Economic and Environmental Pathway of a Shell Farm

63

Ausa–Corno River system in connection with the lagoon (Piani et al., 2005). In 1984, due to application of law rules, a modern wastewater system treatment was installed; hence no more inputs of Hg should have been released into the environment.

Covelli et al. (2009) stated that although uncontrolled Hg discharge from the CAP stopped, the artificial channel connecting the industrial area to the upper river course is still an active Hg source for the Marano lagoon environment.

During the application of the WFD 2000/60/EC, ARPA FVG investigated the distribution of ten elements (Al, As, Cd, Cr, Fe, Hg, Ni, Pb, V and Zn) and of MeHg in surface sediment (0-5 cm) collected in 28 stations distributed throughout the lagoon ecosystem.

In this monitoring plan some critical rose. However, where some elements exceed the environmental quality standards (EQS) set by the WFD it should be stressed that these represent, similarly to what reported for the adjacent Gulf of Trieste (Acquavita et al., 2010b), the values underlying the region's natural lagoon geochemistry.

Previous data for As in the Gulf of Trieste showed a mean value of 7.4 µg g^{-1}. In the lagoon an average of value 8.9±1.7 µg g^{-1} confirm that sediments can be considered as uncontaminated (Francesconi & Edmonds, 1997). Moreover, only one site with 12.5 µg g^{-1} exceeded the EQS (12.0 µg g^{-1}) fixed by the law. For cadmium, the mean concentration detected was equal to 0.17±0.04 µg g^{-1}, with the whole set of data <0.30 µg g^{-1} (EQS). The element was significantly correlated with Fe (r=0.7940, p<0.05) and with Cr, Ni, Pb and Zn, thus suggesting a common origin among these elements. Cr exceeds the EQS (50 µg g^{-1}) only in some stations (mean value 45.6±12.3 µg g^{-1}. The Pearson matrix correlation showed a distribution mainly influenced by mineralogy (Cr vs. Al, r=0.9255, p<0.05) rather than the textural characteristics. A good correlation was also found with Zn and Pb. This latter showed values quite low (16.9±4.7 µg g^{-1}) if compared with those reported by Covelli & Fontolan (1997) mean 55 µg g^{-1} with a maximum of 144 µg g^{-1}. Zn exhibits an extreme variability also in uncontaminated sediments. In the Gulf of Trieste a wide range of values from 8 to 213 µg g^{-1}) is reported (Stefanini, 1971; Ministero dell'Ambiente, Servizio Difesa Mare, 2005). Sediments of the lagoon showed a mean value of 68.5±19.4 µg g^{-1} and the significant correlation with Al (r=0.9003, p<0.05) suggest that there no significant anthropogenic contribution. Ni always exceeds the EQS (30 µg g^{-1}) proposed, thus ARPA FVG will pose particular attention in the future monitoring. The data collected for Hg confirms a range between 0.5 to 8.1 µg g^{-1}, and clearly that Grado is more heavily polluted than Marano. However, some spikes were detected at the Ausa-Corno River mouth (5.9 µg g^{-1}). MeHg contents were similar (2.0±0.9 ng g^{-1}) to those already observed by Covelli et al. (2008). Since the production of methylated Hg species depends on several factors (i. e., redox state of the system, presence and quality of organic carbon, presence/absence of specific bacteria solforiduttori) but not on total Hg contents, no correlation was found between the two species (r=0.4343, p<0.5).

The concentration of Hg, Cd, Pb and benzo(a)pyrene are semiannually detected in bivalves (*T. philippinarum, C. gallina, C. chione* and *M. galloprovincialis*) on the basis of Regulation (EC) 1881/2006, setting maximum levels for certain contaminants in foodstuffs. During the ten years monitoring conducted by ARPA FVG (2001-2010), concentrations of these heavy metals were always below the maximum level. In addition, benzo(a)pyrene levels were always under limit of detection. The maximum level of MeHg in Manila clam from shell farm in the Lagoon of Marano was 0.089 mg/kg wet weight. (Rampazzo et al., 2009).

5. Cultivation of *T. philippinarum* in the Lagoon of Marano and Grado

5.1 Production cycle

The cultivation of T. *philippinarum* in the Lagoon of Marano and Grado is characterized by three phases: production of larvae and seed, pre fattening and fattening. The production of larvae and seed take place in the hatchery and in this phase the conditioning of broodstock occurs through acclimation with controlled temperature and a balanced diet. The emission of gametes is induced by thermal shock and after the fertilization the true phase of cultivation takes place with the growth of larvae up to metamorphosis stage (220 µm). The subsequent growth of the seed occurs in tanks with controlled temperature and by intake of phytoplankton cultures. When the seed reaches a size of 1.5-3 mm the next phase of pre fattening starts. This is a very delicate step and an unsuccessful of the pre fattening could seriously affect the entire cycle of production. The techniques used in the Lagoon of Marano and Grado are carried out in plastic or fiberglass tanks at ground (Fig. 3), on floating upweller or directly on the lagoon bottom, where juveniles are protected by laying of plastic nets (Fig. 4). These plastic nets protect the juvenile clams from predators such as crabs and birds. This step ends when Manila clams reach a size of about 12-15 mm, after this the product is collected, selected, counted and prepared for the next phase called fattening.

Fig. 3. Pre fattening in plastic tanks

The fattening is conducted on the lagoon bottom with pelitic sand sediment and some of the culture areas are periodically emerged during low tide (–40 cm to +10 cm). In this phase the seed of about 15 mm is placed with a density of about 150 individuals/m², for a period of 24 to 36 months. Normally 50% of the juveniles survive until the commercial size (35-40 mm, 15 g)

Manila Clam (Tapes philippinarum Adams & Reeve, 1852) in the Lagoon of Marano and Grado
(Northern Adriatic Sea, Italy): Socio-Economic and Environmental Pathway of a Shell Farm

65

Fig. 4. Laying of plastic nets to protect the juvenile clams from predators

after about three years. In particular during a production cycle the areas devoted to fattening of Manila clams are seeded in an alternate mode; in the second year the growth of molluscs, sanitary and environmental conditions are tested, during the third year mature clams are harvested. Before the subsequent cycle of production the culture beds have to spend a rest period (Zentilin & Orel, 2009).

5.2 Production system

All further steps of the cultivation in the lagoon are conducted with specific equipments and boats: lagoon bottom treatment, seeding, laying and removing of plastic nets (if necessary), cleaning of nets, harvesting. In the past all operations were manually conducted and only after specific experiments, an accurate planning and a lot of experience, peculiar mechanized systems were employed for each phase of the processing in this intertidal environment. Many requirements had to be performed to realize the ideal prototype of operating machine: the choice of resistant materials adapted to operate in the marine environment, the choice of the dimension and a correct distribution of weights, the estimation of the pressure applied to the lagoon sediments and the operating security. These requirements led to perform three models of operating machine: a model of seeder machine combined to remove oyster shells from lagoon bottom and a system to lay plastic nets (Fig. 5); a model for the cleaning and removing of plastic nets and finally a model of shallow hydraulic escalator harvester. The impact of these prototypes, in particular on sediment texture and benthic community, was accurately studied to verify the environmental sustainability of each operation system.

Since 2004 a new multifunctional vessel (M/S ADA) is employed in areas with depth up to -40 cm (Fig. 6). ADA has the following characteristics: 19.34 GT, steel hull, total length 21 m, 7 m in wide and 1.25 m in height. ADA is equipped with a diesel engine of 347 HP and is

Fig. 5. Seeder machine combined to remove oyster shells

Fig. 6. Multifunctional vessel

Manila Clam (Tapes philippinarum Adams & Reeve, 1852) in the Lagoon of Marano and Grado
(Northern Adriatic Sea, Italy): Socio-Economic and Environmental Pathway of a Shell Farm

67

powered by three feet hydraulic power plants; the two feet aft side also act as a rudder. This vessel is also provided by two winches at the bow and one aft, and both bear anchor mooring. The winch-system has the function to advance or retreat the vessel during harvesting operations without the aid of the propellers. This system permits the elimination of sediment resuspension caused by turbulence of propeller on the shallow water. The studies conducted on this kind of vessel, fully demonstrated that such production system constitutes the best compromise for environmental sustainability and economic yield achievement (Zentilin & Orel, 2009).

5.3 Control measures

Regulations (EC) No 853/2004 and No 854/2004 of the European Parliament and of the Council, establish specific rules for the hygiene of foodstuffs and for the organisation of official controls on products of animal origin intended for human consumption respectively. In this way clam farmers may only harvest living bivalve molluscs from production areas with fixed locations and boundaries that the competent authority has classified. The competent authority in fact must fix the location and boundaries of production and relaying areas that it classifies and it may, where appropriate, do so in cooperation with the food business operators. In addition the competent authority must classify production areas from which it authorises the harvesting of living bivalve molluscs as being of one of three categories according to the level of faecal contamination, as following: Class A areas from which living bivalve molluscs may be collected for direct human consumption; Class B areas from which living bivalve molluscs may be collected, but placed on the market for human consumption only after treatment in a purification centre or after relaying so as to meet the health standards, molluscs from these areas must not exceed the limits of a five-tube, three dilution Most Probable Number (MPN) test of 4,600 *Escherichia coli* per 100 g of flesh and intervalvular liquid; finally Class C areas from which living bivalve molluscs may be collected but placed on the market only after relaying over a long period so as to meet the health standards, molluscs from these areas must not exceed the limits of a five-tube, three dilution MPN test of 46,000 *E. coli* per 100 g of flesh and intervalvular liquid.

If the competent authority decides in principle to classify a production or relaying area, it must make an inventory of the sources of pollution of human or animal origin likely to be a source of contamination for the production area; examines the quantities of organic pollutants which are released during the different periods of the year; determines the characteristics of the circulation of pollutants by virtue of current patterns, bathymetry and the tidal cycle in the production area; establishes a sampling programme of bivalve molluscs in the production area which is based on the examination of established data, and with a number of samples, a geographical distribution of the sampling points and a sampling frequency which must ensure that the results of the analysis are as representative as possible for the area considered. Classified relaying and production areas must be periodically monitored to check: there is no malpractice with regard to the origin, provenance and destination of living bivalve molluscs; the microbiological quality of bivalves in relation to the production and relaying areas; the presence of toxin-producing plankton in production and relaying waters and biotoxins in living bivalves and for the presence of chemical contaminants in molluscs. Sampling plans must be drawn up providing for such checks to take place at regular intervals, or on a case-by-case basis if harvesting periods are irregular. The geographical distribution of the sampling points and the sampling frequency must ensure that the results of the analysis are as representative as possible for the area considered. Where the results of sampling show that the health standards for molluscs are exceeded, or that there may be otherwise a risk to human health, the authority must close the production area concerned, preventing the harvesting. The same may re-open a closed production area only if the health standards for molluscs once again comply with Community legislation. To decide on the classification, opening or closure of production areas, the authority may take into account the results of controls that food business operators or organisations representing them have carried out. In that event, the authority must have designated the laboratory carrying out the analysis and, if necessary, sampling and analysis must have taken place in accordance with a protocol that the authority and the operators or organisation concerned have agreed.

The Decision of the Regional Government of Friuli Venezia Giulia (DGR No 124/2010) applies the above mentioned EC Regulations in the marine coastal and transitional waters, in term of classification of harvesting areas and the guidelines for the monitoring protocol of these. In particular Fig. 7 shows that the harvesting areas in the Marano and Grado Lagoon are identified as Class B. In this way the harvested Manila clams need to be treat in a purification centre so as to meet the health standards of three dilution Most Probable Number (MPN) test of <230 *E. coli* per 100 g of flesh and intervalvular liquid.

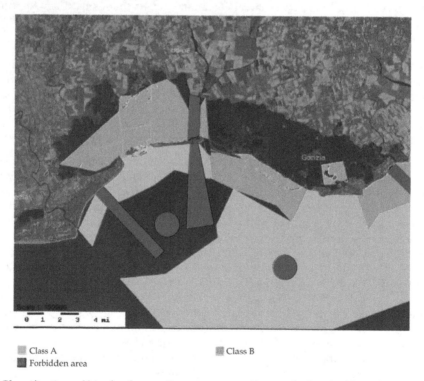

■ Class A   ▓ Class B
■ Forbidden area

Fig. 7. Classification of bivalve harvesting areas according to the level of faecal contamination

Manila Clam (Tapes philippinarum Adams & Reeve, 1852) in the Lagoon of Marano and Grado
(Northern Adriatic Sea, Italy): Socio-Economic and Environmental Pathway of a Shell Farm

69

6. Impact of *T. philippinarum* cultivation on marine ecosystem

The minimization of negative effects, both direct and indirect, of fishing and aquaculture activities is perceived to be an important component of management plans in heavily exploited ecosystems (Benaka, 1999). Demersal resources, such as Manila clam, need to be harvested by use of mobile gears (mainly trawl nets and dredges), producing disturbances which may exceed any other natural and anthropogenic disturbance on the lagoon or the marine ecosystem (Watling & Norse, 1998). Bottom gear may cause widespread physical disturbance in sediments affecting benthic communities, removing both target and non-target species, and altering habitats. Many studies have been carried out to assess the impact of fishing gear used for harvesting molluscs, particularly dredges, on bycatch species and benthic communities (e.g. Pranovi et al., 2001; Hauton et al., 2003; Gaspar et al., 2009; Leitão et al., 2009). The few studies available on the effects of commercial fishing on target species focus mainly on the qualitative/quantitative assessment of the discarded clams and on the selectivity of the fishing gear (Gaspar et al., 2003; Morello et al., 2005; Kraan et al., 2007). In addition the effects of dredging on habitats include changes in the physical structure and chemistry of the environment, sediment suspension and its redistribution (Gaspar & Chicharo, 2007). The alteration of sediment features may influence the colonization and presence of benthic species, including target ones, leading to long-term changes in community structure (Pranovi & Giovanardi, 1994; Pranovi et al., 1998). In this way, there has been increasing recent interest in evaluating the collateral effects of commercial cultivation of bivalves on the marine environment (Sorokin et al., 1999; Bartoli et al., 2001; Jie et al., 2001; Dame et al., 2002), especially on macrobenthic communities (Kaiser et al., 1996; Drake & Arias, 1997; Spencer et al., 1997; Gaspar et al., 2002; 2003). The main effect on non-target species is a reduction in species richness and abundance (Commito, 1987; Dittman, 1990; Guenther, 1996; Ragnarsson & Raffaelli, 1999; Beadman et al., 2004; Pranovi et al., 2004), although in some cases the opposite effect has been recorded (Mantovani et al., 2006). This latter may be caused by the use of plastic nets to protect juvenile clams from predation by shorebirds and crabs (Spencer et al., 1992), which increased sedimentation rates and consequently the density of some species of infaunal deposit-feeding worms (Spencer et al., 1997).

In the Venice Lagoon the harvesting by "rusca" produces a V-shaped furrow (about 60 cm wide and 7 cm deep) and a plume of resuspended sediment with a significant increase (up to two orders of magnitude greater than undisturbed areas) of suspended particulate matter and increased C_{tot}, C_{org}, N_{tot} and sulphide concentrations in the water column. During experiments "rusca" hauls significantly reduced macrofauna density, whereas no significant effect on meiofauna was detected (Pranovi et al., 2004). The resuspension caused by "rusca" fishing activity could be an important factor in determining food quality and quantity available to filter feeders as described by De Jonge & Van Beusekom (1992), for other resuspension sources. All this could explain the *"Tapes paradox"*, which is the apparent benefit of *Tapes* populations to exploitation. As reported by Pranovi et al. (2003), to sustain the huge clam biomass an external energetic input is required, because the concentration of the suspended food in the Venice Lagoon is occasionally below the threshold demanded by Manila clams (Pranovi et al., 2004).

One of the main effects of the resuspension activity is an increase in water turbidity (Black & Parry, 1994), which could profoundly affect primary production (both in the water column and on the bottom). In the Venice Lagoon a significant increase in water turbidity has been recorded since the beginning of the 1990s (Sfriso & Marcomini, 1996). In this way, the resuspended sediment could be transported by natural hydrodynamics, e.g. tidal currents, and eventually reach deeper channels and be driven outside the lagoon. Even if the resuspended sediment is redeposited in other shallow areas, it would not be well stabilized and would therefore be more exposed to erosion processes. So in the Venice Lagoon, resuspension due to mechanical clam harvesting could produce an additional effect on the natural erosion of the shallow bottoms, which is at present one of the major points in the safeguard policy of the Venice Lagoon (Pranovi et al., 2004).

Experimental studies were conducted also in the Lagoon of Marano to test the impact of the "maranese rake", the shallow hydraulic escalator harvester and the hand harvesting on benthic communities and sediment texture and to estimate the regeneration times of macrozoobenthos after the employ of these harvester methods (Orel et al., 2001; 2002; 2005). Everyone leads to a drop of richness and abundance of macrozoobenthos and a loss of fine sediments after the haul of harvesters or the hand method. Nevertheless, after 4-5 months from the experimental harvesting, sedimentary and biological conditions were restored. In this way we could assume that harvesting of Manila clam in the Lagoon of Marano and Grado not involve any relevant environmental modification in the short term and space, but on the contrary the reiteration of the activities in term of space and time could induce long term negative impacts, by progressive reduction of sediment thickness (Orel et al., 2001; 2002).

7. Future approaches and production sustainability

7.1 Trends

Given forecasts of increasing air and sea temperatures (Hulme et al., 2002) it might be expected that the Manila clam will spread to more sites around the coasts of Britain and Europe. The magnitude of the environmental consequences of the naturalisation of the Manila clam at these sites is likely to depend upon the density that the species attains. This in turn is likely to depend upon the future environmental conditions and the intensity of any fishery for this commercially valuable resource. Thank to Regulation (EC) No 708/2007, concerning use of alien and locally absent species in aquaculture, the risk assessment to cultivate Manila clam in the EU is no more necessary. After 25 years from its introduction, this species has never threatened the original ecosystem of the Lagoon of Marano and Grado. On the contrary the development of its sustainable culture could represent a benefit to the local fisheries economy.

7.2 Responsible practices toward sustainable production

In the Lagoon of Marano and Grado a temporary association among business operators was constituted in 2009. This association includes ALMAR Society, Molluschicoltura Maranese Society and the Cooperative of fishermen St. Vito of Marano Lagunare. The goal of this historical cooperation is to start together a pathway towards the sustainable aquaculture of Manila clam in the Lagoon of Marano. After many years of conflicts, the results, the prestige and the reliability got by ALMAR and Molluschicoltura Maranese in the shell farm business, the Cooperative St. Vito, grouping the major part of the fishermen in Marano Lagunare, perceived the necessity to enter the way of the cultivation in the lagoon. Anyway before the constitution of the temporary association, the cooperative already got in 2005 a financial

Manila Clam (Tapes philippinarum Adams & Reeve, 1852) in the Lagoon of Marano and Grado
(Northern Adriatic Sea, Italy): Socio-Economic and Environmental Pathway of a Shell Farm

71

support from EU to realize an hatchery plant sited close to the town. This hatchery has a theoretical production of 200,000,000 ind/year, whereas that in ALMAR is 400,000,000 ind/year. The new hatchery equipped itself of biologists able to follow the entire productive process.

In 2006 a Decision of the Regional Government of Friuli Venezia Giulia approved the new lagoon areas suitable for the fattening of Manila clam, thank to a research conducted by the Department of biology of the University in Trieste in collaboration with ALMAR Society. The study took into consideration: sediment texture, depth profile, hydrology, primary production, macrozoobenthic communities, presence or absence of seagrass and macroalgae, growth rates of clams for each considered zone, microbial suitability in agreement to UE rules, chemical conditions of sediments, anthropogenic pressures, potential disturbs to wild birds, closeness to navigation channels and so on.

At present the total area devoted to Manila clam farming amounts to about 800 hectares (Fig. 8) of which 130 are in full regime of production.

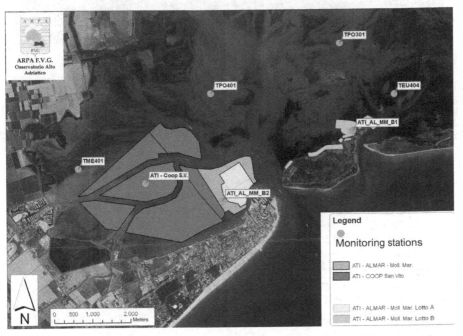

Fig. 8. Areas devoted to Manila clam farm in the Lagoon of Marano and locations of sampling stations for the environmental monitoring

In 2010 a new harvesting gear was introduced in Lagoon of Marano to operate inside the assigned area to Cooperative St. Vito (see ATI – COOP San Vito in Fig. 8). This gear is a small hydraulic clam dredge (Fig. 9), called "idrorasca", originally designed and constructed to be used in the Manila clam natural banks located in the transitional waters of Po River Delta. The use of this new harvesting method was authorized in 2004 after a series of researches inherent the environmental sustainability of this gear in the delta area (Turolla, 2008).

Fig. 9. Hydraulic clam dredge "idrorasca"

In 2008 the use of idrorasca was authorized and tested in the Lagoon of Grado too, but the experiment failed because of the high mortality of Manila clam juveniles in that area. Thus the fishermen of Grado had to give up the first attempt of cultivation in their lagoon. Vice versa the area entrusted to maranese fishermen seems to be very suitable to the growth of clams. In winter 2011 the cooperative has granted to fishermen the use of 30 idrorasca to treat the bottom before the seeding, whereas a plot was already seeded in 2010. Before starting the new gear, a monitoring program was established with the regional agency for the protection of the environment (ARPA FVG), in order to evaluate the sustainability of this new method of harvesting in the Lagoon of Marano during a whole production cycle. In this way seven monitoring stations were selected (Fig. 8) to check the status of macrozoobenthos, sediment texture, chemical pollutants in clam tissues, both inside and around farm areas. Of course in this initial phase the fishermen are strictly watched to prevent any accident or illegal harvesting out of areas. They have to use the gear strictly following the procedures established by the regional authority: max speed of tow 1 km/h, pressure of water flux must not exceed 3.5 atm, no more than two hauls in each pathway and over all they must operate during the high slack tide, to prevent as much as possible the loss of fine sediments.

The environmental and socio economic vicissitudes of the Manila clam in the Lagoon of Marano and Grado could represent a case study: after more than 20 years of fights, farmers, fishermen and environmental policy seem to have found the pathway toward a sustainable shell farm. *T. philippinarum* arrived in the northern Adriatic lagoons as an alien species in '80th, leading to a total derangement of socio economic and environmental conditions. Now after almost 30 years, thanks to the model of shell farming in the Lagoon of Marano, this acclimated species could constitute strength for the fisheries economy of the lagoons and the northern Adriatic too.

Manila Clam (Tapes philippinarum Adams & Reeve, 1852) in the Lagoon of Marano and Grado
(Northern Adriatic Sea, Italy): Socio-Economic and Environmental Pathway of a Shell Farm

73

8. Aknowledgements

Authors are grateful to A. D'Aietti and P. Rossin for the maps and in particular to G. Orel e G. Mattassi for their efforts over the years towards a sustainable aquaculture in the Lagoon.

9. References

Acquavita, A.; Celio, M.; Mattassi, G.; Predonzani, S., & Rossin, P. (2010a). Vulnerability of the lagoon ecosystem: a case study from the Marano and Grado Lagoon. *4th Lagunet- Marsala (Tp)*, 27-30 ottobre

Acquavita, A.; Predonzani, S.; Mattassi, G.; Rossin, P.; Tamberlich, F.; Falomo, J.; & Valic, I. (2010b). Heavy Metal Contents and Distribution in Coastal Sediments of the Gulf of Trieste (Northern Adriatic Sea, Italy). *Water, Air and Soil Pollution*, No. 211, pp. 95-111

ARPA-FVG (2006). Caratterizzazione chimica dell'areale marino costiero della Regione Friuli- Venezia Giulia – *Stato di fatto e ricerca dei valori di fondo. 1° rapporto sui sedimenti pelitici*

Bardach, J.E.; Ryther, J.H. & McLarney, W.O. (1972). Aquaculture. The farming and husbandry of freshwater and marine organisms. John Wiley and Sons, New York. P. 868

Bartoli, M., Nizzoli, D., Viaroli, P., Turolla, E., Castadelli, G., Fano, E. A., & Rossi, R. (2001). Impact of *Tapes philippinarum* farming on nutrient dynamics and benthic respiration in the Sacca di Goro. *Hydrobiologia*, No. 455, pp. 203-212

Beadman, H. A., Kaiser, M. J., Galandi, M., Shucksmith, R., & Willows, R. I. (2004). Changes in species richness with stocking density of marine bivalves. *Journal of Applied Ecology*, No. 41, pp. 464-475

Benaka, L. (1999). Fish habitat: essential fish habitat and rehabilitation. *American Fishery Society*, Bethesda

Black, K. P. & Parry, G. D. (1994). Sediment transport rates and sediment disturbance due to scallop dredging in Port Philip Bay. *Memoirs of the Queensland Museum*, No. 36, pp. 327-341

Bosa, S., & Petti, M. (2004). A finite volume model of flow and sediment transport in the lagoon of Grado and Marano In: Greco, Carravetta, Della Morte (Eds.), *River Flow 2004*. Taylor & Fracis Group, London, UK, pp. 677-687

Bourne, N. (1982). Distribution, reproduction and growth of Manila clam, *Tapes philippinarum* (Adams and Reeves), in British Columbia. *J. Shellfish Res.* 2: 47-54

Brambati, A. (1970). Provenienza, trasporto e accumulo dei sedimenti recenti nelle lagune di Marano e di Grado e nei litorali tra i fiumi Isonzo e Tagliamento. *Bollettino della Societa` Geologica Italiana*, No. 9, pp. 281–329

Brambati, A. (1972). Clay mineral investigation in the Marano and Grado Lagoons (northern Adriatic Sea). *Bollettino della Societa` Geologica Italiana*, No. 91, pp. 315–323

Brambati, A. (2001). Coastal sediments and biota as indicators of Hg contamination in the Marano and Grado Lagoons. *RMZ – Materials & Geoenvironment* No. 48, pp. 165–171

Breber, P. (1985). L'introduzione e l'allevamento in Italia dell'arsella del pacifico *Tapes semidecussatus* Reeve (Bivalvae veniridae). *Oebalia* vol. IX, N°2, pp 675-680

Cesari, P. & Pellizzato, M. (1985). Microflora associated to outbreaks of « Brown Ring» from clams (*Tapes semidecussatus*) cultured in southwestem Spain. In: A. Figueras (Editor), *4th Int. Colloq. on Pathology in Marine Aquaculture*. CSIC, Vigo, Spain, 56 p

Chew, K. (1989). Manila clam biology and fishery development in western North America. In "Clam mariculture in North America". Development in *Aquaculture and Fisheries Science*, 19. Ed. by J.J. Manzi and M. Castagna (Elsevier, Amsterdam): pp. 243-261.

Commito, J. A. (1987). Adult-larval interactions: predictions, mussels and cocoons. *Estuarine. Coastal and Shelf Science*, No. 25, pp. 599-606

Cottiglia, M.; Marsal, M., & Tagliasacchi, L. (1988). Esperienze di allevamento di *Tapes philippinarum* in Sardegna. *Quad. ist. Idrobiol Acquacolt "G Brunelli"* vol. 8 pp 3-77

Covelli, S., & Fontolan, G. (1997). Application of a normalization procedure in determining regional geochemical baselines, Gulf of Trieste, Italy. *Environmental Geology*, No. 30(1/2), pp. 34-45

Covelli, S., Faganeli, J., Horvat, M., & Brambati, A. (2001). Mercury contamination of coastal sediments as the result of long-term cinnabar activity (Gulf of Trieste, northern Adriatic Sea). *Applied Geochemistry*, No. 16, pp. 541-558

Covelli, S.; Acquavita, A.; Piani, R.; Predonzani, S., & De Vittor, C. (2009). Recent contamination of mercury in an estuarine environment (Marano lagoon, Northern Adriatic, Italy). *Estuarine, Coastal and Shelf Sciences*, No. 82, pp. 273-284

Covelli, S.; Faganeli, J.; De Vittor, C.; Predonzani, S.; Acquavita, A., & Horvat, M. (2008). Benthic fluxes of mercury species in a lagoon environment (Grado lagoon, Northern Adriatic Sea, Italy). *Applied Geochemistry*, No. 23, pp. 529-546

Covelli, S.; Piani, R.; Acquavita, A.; Predonzani, S., & Faganeli, J. (2007). Transport and dispersion of particulate Hg associated to a river plume in coastal Northern Adriatic environments. *Marine Pollution Bulletin* No. 55, pp. 436-450

Da Ros, L.; Moschino, V.; Meneghetti, F., & Marin, M. G. (2005). Aspetti del ciclo riproduttivo e del metabolismo energetico di *Tapes Philippinarum* in Laguna di Venezia. In: *La filiera della vongola – Tapes philippinarum in Italia*, Boatto, V., & Pellizzato, M., pp. 77-88, Franco Angeli

Dame, R.; Bushek, D.; Allen, D.; Lewitus, A.; Edwards, D.; Koepfler, E., & Gregory, L. (2002). Ecosystem response to bivalve density reduction: management implications. *Aquatic Ecology*, No. 36, pp. 51-65

De Jonge, V.N., & van Beusekom, J.E.E. (1992). Contribution of resuspended microphytobenthos to total phytoplankton in the Ems estuary and its possible role for grazers. *Netherlands Journal of Sea Research*, No. 30, pp. 91-105

Devauchelle, N. (1990). Sexual development and maturity of *Tapes Philippinarum. Tapes philippinarum*, biology and experimentation. ESAV. Ed. Verone. pp. 48-62.

Di Marco, P.; Lombardi, L., & Rambaldi , E. (1990). Allevamento sperimentale della vongola verace *Tapes philippinarum* nel lago di Sabaudia *Quad. ist. Idrobiol Acquacolt "G Brunelli"* vol. 10 pp 15-32

Dittman, S. (1990). Mussel beds – amensalism or amelioration for intertidal fauna? *Helgoland Marine Research*, No. 44, pp. 335-352

Dorigo, L. (1965). The Lagoon of Grado and Marano and its Mouths. Researches and Hydraulic Measurements. Tech. rep., Magistrato alle Acque, Ufficio Idrografico. Grafiche Gasparoni, Venezia

Drake, P., & Arias, A. M. (1997). The effect of aquaculture practices on the benthic macroinvertebrate community of a lagoon system in the Bay of Cadiz (Southern Spain). *Estuaries*, No. 20, pp. 677-688

European Union, 2000. Directive 2000/60/EC of the European Parliament and of the Council of 23 October 2000 establishing a framework for Community action in the

Manila Clam (Tapes philippinarum Adams & Reeve, 1852) in the Lagoon of Marano and Grado
(Northern Adriatic Sea, Italy): Socio-Economic and Environmental Pathway of a Shell Farm

75

field of water policy (Water Framework Directive). *Official Journal of the European Communities C*, L 327 22/12/2000

Eversole, A.G. (1989). Gametogenesis and spawning in North American clam populations: implications for culture: In: J.J. Manzi & M. Castagna (eds.). *Clam mariculture in North America*. Elsevier Science Publishers B.V., Amsterdam. pp. 75-109

FAO (2011). Available from www.fao.org

Ferrarin, C.; Umgiesser, G.; Bejo, M.; Bellafiore, D.; De Pascalis, F.; Ghezzo, M.; Mattassi, G. & Scroccaro, I. (2010). Hydraulic zonation of the lagoons of Marano and Grado, Italy. A modelling approach. *Estuarine, Coastal and Shelf Sciences*, No. 87, pp. 561-572

Fontolan, G.; Pillon, S.; & Facchin, G. (2009). Multidecadal salt marsh evolution in the northern Adriatic lagoons, Italy: erosional style and morphological adaptation to transgressive forcing (In Italian; Trasformazioni territoriali ed aspetti fisici di adattamento dell'ecosistema naturale delle lagune di Grado e Marano conseguenti all'innalzamento del livello del mare). VECTOR - II Yearly Workshop (Roma, 25-26 February 2009), http://vector.conismamibi.it/files/II%20workshop/orali/Fontolan.pdf

Francesconi, K.A. & Edmonds, J.S. (1997). Arsenic and marine organisms. *Advances in Inorganic Chemistry*, No. 44, pp. 147-189

Gaspar, M. B. & Chicharo, L. (2007). Modifying dredges to reduce by-catch and impacts on the benthos. In: *By-catch reduction in the world's fisheries*, Kennelly S. J., pp. 95-140, Springer

Gaspar, M. B., Leitão, F., Santos, M. N., Sobral, M., Chìcharo, L., Chìcharo, A., & Monteiro, C. C. (2003). Size selectivity of the *Spisula solida* dredge in relation to tooth spacing and mesh size. *Fisheries Research*, No. 60, pp. 561-568

Gaspar, M. B.; Leitão, F.; Santos, M. N.; Sobral, M.; Chìcharo, L.; Chìcharo, A. & Monteiro, C. C. (2002). Influence of mesh size and tooth spacing on the proportion of damaged organisms in the catches of the Portuguese clam dredge fishery. *ICES Journal of Marine Science*, No. 59, pp. 1228-1236

Gaspar, M.B.; Carvalho, S.; Costantino, R.; Tata-Regala, J.; Cúrdia, J.; & Monteiro, C. C. (2009). Can we infer dredge fishing effort based on macrobenthic community structure? *ICESJournal of Marine Science*, No. 66, pp. 2121-2132

Giorgiutti, E.; Libralato, M. & Pellizzato, M. (1999). Sperimentazioni di acquicoltura in laguna di Caorle (Ve). A.S.A.P., 29 p.

Goulletquer, P. (1997). A bibliography of Manila clam *Tapes philippinarum* IFREMER RIDRV-97.02/RA/La Trembalde, 122p.

Guenther, C. P. (1996). Development of small *Mytilus* beds and its effect on resident intertidal macrofauna. *Marine Ecology*, No. 17, pp. 117-130

Hauton, C.; Hall-Spencer, J.M. & Moore, P.G. (2003). An experimental study of the ecological impacts of hydraulic bivalve dredging on maerl. *ICES Journal of Marine Science*, No. 60, pp. 381-392

Hulme, M.; Jenkins, G.; Lu, X.; Turnpenny, J.; Mitchell, T.; Jones, R.; Lowe, J.; Murphy, J.; Hassell, D.; Boorman, P.; McDonald, R. & Hill, S. (2002) *Climate Change Scenarios for theUnited Kingdom: The UKCIP02 Scientifi c Report*. Norwich: Tyndall Centre for Climate Change Research, School of Environmental Sciences, University of East Anglia

Jie, H.; Zhinan, Z.; Zishan, Y. & Widdows, J. (2001). Differences in the benthic –pelagic particle flux (biodeposition and sediment erosion) at intertidal sites with and without clam (*Ruditapes philippinarum*) cultivation in eastern China. *Journal of Experimental Marine Biology and Ecology*, No. 261, pp. 245-261

Kaiser, M. J.; Edwards, D.B. & Spencer, B. E. (1996). Infaunal community changes as a result of commercial clam cultivation and harvesting. *Aquatic Living Resources*, No. 9, pp. 57-63

Kraan, C.; Piersma, T.; Dekinga, A.; Koolhaas, A. & Van der Meer, J. (2007). Dredging for edible cockles (*Cerastoderma edule*) on intertidal flats: short–term consequences of fisher patch-choise decisions for target and non-target benthic fauna. *ICES Journal of Marine Science*, No. 64, pp. 1735-1742

Leitão, F.; Gaspar, M. B.; Santos, M. N. & Monteiro, C.C. (2009). A comparison of bycatch and discard mortality in three types of dredge used in the Portuguese *Spisula solida* (solid surf clam) fishery. *Aquatic Living Resources*, No. 22, pp. 1-10

Leoni, L. & Sartori, F. (1997). Heavy metal and arsenic distributions in sediments of the Elba-Argentario basin, southern Tuscany, Italy. *Environmental Geology*, No. 32, pp. 83-92

Magoon, C. & Vining, R. (1981). Introduction to shellfish aquaculture. Washington *Dept. of Natural Resources*, Seattle, 28 p.

Mantovani, S.; Castaldelli, G.; Rossi, R. & Fano, E.A. (2006). The infaunal community in experimental seeded low and high density Manila clam (*Tapes philippinarum*) beds in a Po River Delta lagoon (Italy). *ICES Journal of Marine Science*, No. 63, pp. 860-866

Marocco, R. (1995). Sediment distribution and dispersal in northern Adriatic lagoons (Marano and Grado paralic system). *Geologia*, No. 57 (12), pp. 77- 89

Mattassi, G.; Daris, F.; Nedoclan, G.; Crevatin, E.; Modonutti, G.B. & Lach, S. (1991). La qualità delle acque della Laguna di Marano. Regione Autonoma Friuli Venezia Giulia – U.S.L. n.8 "Bassa Friulana", pp. 101

Menchini, G.; Amoroso, C.; Barbanti, A.; Ramieri, E.; Melli, F.; Lopez y Royo F.; Molinari, L.; Simeoni, P. & Bressan, E. (2009). The remediation project of the industrial area of Torviscosa (In Italian; Il progetto di bonifica del sito industriale di Torviscosa). *Rassegna Tecnica del Friuli Venezia Giulia*, Anno LX, Novembre – Dicembre 2009

Milia, M. (1990). Venericoltura in Laguna di Caleri ed in Sacca degli Scardovari. In: *Tapes philippinarum, biologia e sperimentazione*, E.S.A.V. pp. 209–211

Ministero dell'Ambiente e del territorio-Direzione per la protezione della natura (2005). Programma di monitoraggio dell'ambiente marino costiero. Rapporti 2001-2005, Direzione regionale dell'ambiente, ARPA FVG

Morello, E.B.; Froglia, C.; Atkinson, R.J.A. & Moore, P. G. (2005). Hydraulic dredge discards of the clam (*Chamelea gallina*) fishery in the western Adriatic Sea, Italy. *Fisheries Research*, No. 76, pp. 430-444

Orel, G.; Boatto, V.; Sfriso, A. & Pellizzato, M. (2000). Piano per la gestione delle risorse alieutiche delle lagune della Provincia di Venezia. Provincia di Venezia, Ass. Pesca (ed.), Venezia, Italy, 102 pp.

Orel, G.; Zentilin, A.; Zamboni, R.; Grimm, F. & Pessa, G. (2001). Evoluzione delle produzioni ed impatto di alcuni sistemi di raccolta e di pesca di *Tapes philippinarum* (Adams e Reeve, 1850) in uso nella Laguna di Marano (Adriatico Settentrionale). *Biologia Marina Mediterranea*, Vol. 1, No. 8, pp. 432-440

Orel, G.; Fontolan, G.; Burla, I.; Zamboni, R.; Zentilin, A. & Pessa, G. (2002). Aspetti dell'impatto della pesca della vongola verace filippina (*Tapes philippinarum*) con draghe al traino nella Laguna di Marano (Adriatico Settentrionale). *Biologia Marina Mediterranea*, Vol. 1, No. 9, pp. 129-137

Manila Clam (Tapes philippinarum Adams & Reeve, 1852) in the Lagoon of Marano and Grado
(Northern Adriatic Sea, Italy): Socio-Economic and Environmental Pathway of a Shell Farm

77

Orel, G.; Zamboni, R. & Zentilin, A. (2005). Impatto della pesca e della coltura di *Tapes philippinarum* sui fondali delle lagune alto adriatiche. In: *La filiera della vongola – Tapes philippinarum in Italia*, Boatto, V., & Pellizzato, M., pp. 45-57, Franco Angeli

Paesanti, F. (1990). Venericoltura in Sacca di Goro *Tapes philippinarum, biologia e sperimentazione*, E.S.A.V. pp. 212–217

Pellizzato, M.; Penzo, P.; Galvan, T. & Bressan, M. (2005). Insediamento e reclutamento di *Tapes philippinarum*. In: *La filiera della vongola – Tapes philippinarum in Italia*, Boatto, V., & Pellizzato, M., pp. 89-99, Franco Angeli

Petti, M. & Bosa, S. (2004). Pollution transport in the lagoon of Grado and Marano: a two dimensional modelling approach. In: Greco, Carravetta, Della Morte (Eds.), *River Flow* 2004. Taylor & Fracis Group, London, UK, pp. 1183-1192

Piani, R. & Covelli, S. (2000). Contributo antropico di metalli pesanti e ^{137}Cs nei sedimenti del bacino di Buso (Laguna di Marano e Grado, Italia settentrionale). *Studi Trentini di Scienze Naturali-Acta Geologica*, No. 77, pp. 169–177

Piani, R.; Covelli, S; & Biester, H. (2005). Mercury contamination in Marano Lagoon (Northern Adriatic sea, Italy): source identification by analyses of Hg phases. *Applied Geochemistry*, No. 20, pp. 1546–1559

Pranovi, F., & Giovanardi, O. (1994). The impact of hydraulic dredging for short-necked clams, *Tapes* spp., on an infaunal community in the Lagoon of Venice. *Scientia Marina*, No. 58, pp. 345-353

Pranovi, F.; Giovanardi, O. & Franceschini, G. (1998). Recolonization dynamics in areas disturbed by bottom fishing gears. *Hydrobiologia*, No. 375/376, pp. 125-135

Pranovi, F.; Raicevich, S.; Franceschini, G.; Torricelli, P. & Giovanardi, O. (2001). Discard analysis and damage to non-target species in the "rapido" trawl fishery. *Marine Biology*, No. 139, pp. 863-875

Pranovi, F.; Libralato, S.; Raicevich, S.; Granzotto, A.; Pastres, R. & Giovanardi, O. (2003). Mechanical clam dredging in Venice Lagoon: ecosystem effects evaluated with a trophic mass-balance model. *Marine Biology*, No. 143, pp. 393-403

Pranovi, F.; Da Ponte, F.; Raicevich, S. & Giovanardi, O. (2004). A multidisciplinar study of the immediate effects of mechanical clam harvesting in the Venice Lagoon. *ICES Journal of Marine Science*, No. 61, pp. 43-52

Ponurovski S.K. (2008). Population Structure and Growth of the Japanese Littleneck Clam *Ruditapes philippinarum* in Amursky Bay, Sea of Japan *ISSN 1063-0740, Russian Journal of Marine Biology, 2008*, Vol. 34, No. 5, pp. 329–332.

Ragnarsson, S.A. & Raffaelli, D. (1999). Effects of the mussel *Mytilus edulis* on the invertebrate fauna of sediments. *Journal of Experimental Marine Biology and Ecology*, No. 241, pp. 31-43

Rampazzo, F.; Maggi, C.; Bianchi, J.; Emili, A.; Mao, A.; Covelli, S. & Giani, M. (2009). Mercury bioaccumulation in natural and cultured Manila clams (*Tapes philippinarum*) of the Grado and Marano Lagoon (Italy). *Proceedings of 9th International Conference on Mercury as a Global Pollutant*, Guiyang, China, June 2009

Ret, M. (2006). Bilancio idrologico e circolazione idrica della Laguna di Marano e Grado. Master's thesis, Faculty of Engineering, University of Udine, Italy, in Italian

Rigollet, V.; Sfriso, A.; Marcomini, A. & Casabianca, M.L.D. (2004). Seasonal evolution of heavy metal concentrations in the surface sediments of two Mediterranean *Zostera marina* L. beds at Thau lagoon (France) and Venice lagoon (Italy). *Bioresource Technology*, No. 95, pp. 159-167

Scarlato O.A. (1981). Bivalves of temperate waters of the Northwestern part of the Pacifie ocean. Nauka Press, Leningrad, 408 p

Sfriso, A. & Marcomini, A. (1996). Decline of *Ulva* growth in the Lagoon of Venice. *Bioresource Technology*, No. 58, pp. 299-307

Shpigel, M. & Fridman, R. (1990). Propagation of the Manila clam (*Tapes semidecussata*) in the effluent of fish aquaculture ponds in *Eilat, Israel. Aquaculture*, vol. 90. pp. 113-122 SHYFEM http://www.ve.ismar.cnr.it/shyfem

Solidoro, C.; Melaku Canu, D.; Cucco, A. & Umgiesser, G. (2004). A partition of the Venice lagoon based on physical properties and analysis of general circulation. *Journal of Marine Systems*, No. 51, pp. 147-160

Solomons, W., & Föerstner, U. (1984). Metals in the hydrocycle. Springer-Verlag, GmbH, pp. 349

Sorokin, I.I.; Giovanardi, O.; Pranovi, F. & Sorokin, P.I. (1999). Need for restricting bivalve culture in the southern basin of the Lagoon of Venice. *Hydrobiologia*, No. 400, pp. 141-148

Spencer, B.E.; Edwards, D.B. & Millican, P.F. (1992). Protecting Manila clam (*Tapes philippinarum*) beds with plastic netting. *Aquaculture*, No. 105, pp. 251-268

Spencer, B.E.; Kaiser, M.J. & Edwards, D.B. (1997). Ecological effects of intertidal Manila clam cultivation: observations at the end of the cultivation phase. *Journal of Applied Ecology*, No. 34, pp. 444-452

Stefanini, S. (1971). Distribuzione del Li, Na, K, Mg, Ca, Sr, Cr, Mn, Fe, Ni, C, Zn, Pb, C e N organici nei sedimenti superficiali delle Lagune di Marano e Grado (Adriatico Settentrionale). *Studi Trentini di Scienze Naturali, Sez A*, No. 48, pp. 39-79

Turolla, E. (2008). La venericoltura in Italia. In: A. Lovatelli, A. Farías e I. Uriarte (eds). Estado actual del cultivo y manejo de moluscos bivalvos y su proyección futura: factores que afectan su sustentabilidad en América Latina. Taller Técnico Regional de la FAO. 20–24 de agosto de 2007, Puerto Montt, Chile. *FAO Actas de Pesca y Acuicultura*. No. 12. Roma, FAO. pp. 177–188

Umgiesser, G. (1997). Modelling the Venice lagoon. *International Journal of Salt Lake Research*, No. 6, 175-199

Umgiesser, G.; Melaku Canu, D.; Cucco, A. & Solidoro, C. (2004). A finite element model for the Venice Lagoon. Development, set up, calibration and validation. *Journal of Marine Systems*, No. 51, pp. 123-145

Umgiesser, G.; Melaku Canu, D.; Solidoro, C. & Ambrose, R. (2003). A finite element ecological model: a first application to the Venice Lagoon. *Environmental Modelling & Software*, No. 18, pp. 131-145

Watling, L. & Norse, E.A. (1998). Disturbance of the seabed by mobile fishing gear: a comparison to forest clear cutting. *Conservation Biology*, No. 12, pp. 1180-1197

Zentilin, A. (1987). L'allevamento della vongola verace nella laguna di Marano (UD). *Atti della Seconda Giornata della acquacoltura lagunare*, Marano Lagunare, 31 ottobre 1987

Zentilin, A.; Orel, G. & Zamboni, R. (2007). L'introduzione in Europa di Tapes philippinarum (Adams & reeve, 1852), la vongola verace filippina. *Annales Series Historia Naturalis*, Vol. 2, No. 17, pp. 227-232

Zentilin, A.; Pellizzato, M.; Rossetti, E. & Turolla, E. (2008). La venericoltura in Italia a 25 anni dal suo esordio. *Il Pesce*, No. 3, pp. 31-40

Zentilin, A. & Orel, G. (2009). La Laguna di Marano e Grado (Nord Adriatico) e le sue attività di pesca, di molluschicoltura e di vallicoltura. *Hydrores*, No. 31, pp. 210-217

5

The Investigation of the Hydrodynamics of an Artificial Reef

Yan Liu[1], Guohai Dong[1], Yunpeng Zhao[1],
Changtao Guan[2,3] and Yucheng Li[1]
[1]Dalian University of Technology
[2]Yellow Sea Fisheries Research Institute
[3]Key Laboratory of Fishery Equipment and Engineering
China

1. Introduction

Marine resources have been declining all over the world in recent decades. The natural reefs of the world are experiencing higher use and pressures, resulting in anthropogenic influences that are deteriorating coral reefs and causing poor water quality. Artificial reefs are manmade structures that are placed on the seabed deliberately to mimic some characteristics of a natural reef. **(Jensen, 1998)**. Both Japan and the United States have been using artificial reefs for at least 200 years. The first artificial reef was deployed in Japan in the 1700s, and the primary goal was to increase fish-catch. Nowadays, artificial reefs in Japan involve diversified types, integrated materials, complicated structures and large-scale design. The first documented artificial reef in the United States dates from 1830, when logs and rocks were sunk off the coast of South Carolina to improve fishing **(Williams, 2006)**. In many countries, artificial reefs have become important elements in the plans for integrated fishery management. Diverse shapes for artificial reefs, such as cube reefs, circular reefs, cross-shaped reefs and so on, provide favorable circumstances for fish perching, foraging, breeding and defense from enemies. Most artificial constructions in the marine environment consist of a variety of non-natural materials. At present, concrete is most commonly employed as a reef material, including cubes, blocks and pipes. Concrete has also been used in combination with other reef materials such as steel, quarry rock, tires and plastic **(Baine, 2001)**. Many of the world's largest artificial reefs have been deployed as part of a national fisheries program in Japan, where large steel and concrete frameworks have been carefully designed to withstand strong ocean currents **(Seaman, 2007)**. Other popular reef building materials include natural stone and rock, the latter constructed from wide-ranging materials including canvas, anchor blocks, and others. The different faunal assemblages associated with artificial reefs are determined by the physical and chemical nature of the reef materials. **Walker, et al (2002)** quantitatively compared the fish abundance, fish species richness and fish biomass on artificial reef modules constructed of various materials.

There are several proposed mechanisms to explain how artificial reefs increase total biomass production. First, an artificial reef can provide additional food and increase feeding efficiency, and many studies have reported observations of fishes feeding at artificial reefs.

Secondly, affording shelter from predation implies higher survival at artificial reefs than in natural habitats, which is supported by some early experimental studies. **Hixon & Beets, (1989)** tested two corollaries of the limited shelter hypothesis by conducting the experiments in Perseverance Bay, St. Thomas, U.S. Virgin Islands. They suggested that artificial reefs designed for persistent fisheries should include both small holes for small fishes (as refuges from predation) and large holes for predatory species (as home sites). **Leitao, et al (2008)** studied the effect of predation on artificial reef juvenile demersal fish species. The results indicated that interspecific interactions (predator–prey) are important for conservation and management and for the evaluation of the long-term effects of reef deployment. Third, artificial reefs increase the production of natural reef environments by creating vacated space. **Einbinder, et al. (2006)** experimentally tested whether artificial reefs change grazing patterns in their surrounding environment. The results suggest that herbivorous fishes are attracted to the artificial reefs, creating a zone of increased grazing. It is necessary to consider their overall influence on their natural surroundings when planning deployment of artificial reefs. Fourth, artificial reefs have been demonstrated a potential tool for the restoration of marine habitats. The proper role of artificial habitats in aquatic systems continues to be a topic of debate among ecological engineers. A detailed description of the role of artificial reef habitats in the restoration of degraded marine systems has been given by **Seaman, (2007)**. Finally, an important aspect of artificial reefs includes the flow pattern effect. Recent studies have demonstrated the value of investigating the flow field effect in and around artificial reefs as a means of identifying their ecological effect on the proliferation of fishery resources **(Lin & Zhang, 2006)**. When an artificial reef is deployed at the bottom of the sea, many kinds of flow patterns characterized by intense velocity gradients, flow turbulence, vortices, and so on are aroused and develop depending on the shapes, sizes and the different arrangements of artificial reefs. **Zhang & Sun, (2001)** discovered that considerable local upwelling current fields and eddy current fields are generated at the front and the back of artificial reefs, respectively. Upwelling enhances biological productivity, which feeds fisheries, and it is richer in nutrients. Meanwhile, a geometric shaded area is distributed within and around the reef. Consequently, algae and plankton thrive, and therefore, the reefs not only provide a shelter from predators but also an abundant food source.

In view of the above, flow field research is key to the research of artificial reef eco-efficiency. **Liu, et al. (2009)** simulated the flow fields around solid artificial reefs models with different shapes, including cubes, pyramids and triangular prisms, by mean of wind tunnel experiments and numerical simulation methods. **Su, et al. (2007)** conducted particle image velocimetry (PIV) measurements to study the flow patterns within and around an artificial reef. The PIV results were then used to verify the numerical results obtained from finite volume method (FVM) simulations. In engineering practice, the stability of artificial reefs is an important issue in preventing the failure of reef units due to current action. The stability of reefs and sediment erosion on the bottom of artificial reefs rely on the interactions among the current-bottom material-reef system. In this study, the flow field patterns around a hollow cube artificial reef were investigated by employing FVM simulation with unstructured tetrahedral grids for solution of three-dimensional incompressible Reynolds-averaged Navier-Stokes equations. A renormalization group (RNG) k-ε turbulence model was embedded in the Navier-Stokes solver. The pressure and velocity coupling was solved with the SIMPLEC algorithm at each time step. To validate simulation predictions, noninvasive PIV measurements were conducted to measure the flow patterns around the artificial reef. On the basis of the numerical model validation, the flow fields of hollow cube artificial reefs with different altitudes were analyzed using the numerical model. The effects of spacing on the flow field around the parallel and vertical two hollow cube artificial reefs were also analyzed in detail.

2. Numerical methods

Computational fluid dynamics (CFD) are generally used by engineers to solve fluid dynamics problems that involve solving some form of the Navier-Stokes equations. The numerical simulation analysis of flow fields around artificial reef is based on a dynamic, full 3D model elaborated with the aid of FLUENT 6.3 commercial code. The finite volume method are used in the study to solve the three-dimensional Reynolds-averaged Navier-Stokes (RANS) equations for incompressible flows.

2.1 Hypotheses

1.　Incompressible, viscous, Newtonian fluid for the water.
2.　Isothermal flows, regardless of heat exchange in the water.
3.　Flow is in the non-steady state.
4.　The water surface is modeled as a "moving wall" with zero shear force and the same speed as the incoming fluid.

2.2 Hydrodynamic equations solved

The equations solved are the momentum equation, also known as the Navier-Stokes equation, and the continuity equation. This approach is useful for solving laminar flows, but, for turbulent flows, the direct solving of all the vortices in a turbulent flow is expensive. Therefore, a model for turbulence must be added. There are several types of turbulence models available. The most common are the RANS models and large eddy simulation models. In the present study, the two-equation RANS models were chosen to resolve the ensemble averaged flow and model the effect of the turbulent eddies. In the RANS equations, various instantaneous physical parameters are replaced by the time-averaged value. Incompressible flow is assumed such that the equations are as follows:

Momentum equation

$$\rho \frac{\partial}{\partial x_j}(u_i u_j) = -\frac{\partial P}{\partial x_i} + \frac{\partial}{\partial x_j}(\mu \frac{\partial u_i}{\partial x_j} - \rho \overline{u_i' u_j'}) + S_i \tag{1}$$

Continuity equation

$$\frac{\partial u}{\partial x} + \frac{\partial v}{\partial y} + \frac{\partial w}{\partial z} = 0 \tag{2}$$

For Eq. (1), ρ is the mass density; u_i is the average velocity component for x, y, z; P is a body of fluid pressure on the micro-volume; μ is the viscosity; u' is the fluctuation velocity; and i, j=1, 2, 3 (x, y, z). S_i is the source item.

In eddy viscosity turbulence models, the Reynolds stresses are linked to the velocity gradients via the turbulent viscosity, and this relation is called the Boussinesq assumption, where the Reynolds stress tensor in the time averaged Navier-Stokes equation is replaced by the turbulent viscosity multiplied by the velocity gradients. The two-equation models assume an eddy-viscosity relationship for the Reynolds stresses in Eq. (1) given by

$$-\rho \overline{u_i' u_j'} = \mu_t \left(\frac{\partial u_i}{\partial x_j} + \frac{\partial u_j}{\partial x_i} \right) - \frac{2}{3} \rho k \delta_{ij} \tag{3}$$

where μ_t is the eddy viscosity and δ_{ij} is the Kronecker delta.

The k–ε two equations model has become the workhorse for practical engineering modeling of turbulent flows due to its robustness, economy and reasonable accuracy for a wide range of flows. In the derivation of the k–ε model, it is assumed that the flow is fully turbulent and that the effects of molecular viscosity are negligible. Therefore, the model is only valid for fully turbulent flows. An improved version of the k–ε turbulence model, the renormalization group (RNG) k–ε model, is adopted here, which is derived from the transient Navier-Stokes equation and employs a new mathematical method called a "renormalization group". RNG k–ε model provides an option to account for the effects of swirl or rotation by modifying the turbulent viscosity appropriately

The new turbulent equations for "k" and "ε" are introduced as follows.

k equation

$$\frac{\partial}{\partial t}(\rho k) + \frac{\partial}{\partial x_i}(\rho k \overline{u_i}) = \frac{\partial}{\partial x_j}\left(\alpha_k \mu_{eff} \frac{\partial k}{\partial x_j} \right) + G_k - \rho \varepsilon \tag{4}$$

ε equation

$$\frac{\partial}{\partial t}(\rho \varepsilon) + \frac{\partial}{\partial x_i}(\rho \varepsilon \overline{u_i}) = \frac{\partial}{\partial x_j}\left(\alpha_\varepsilon \mu_{eff} \frac{\partial \varepsilon}{\partial x_j} \right) + C_{1\varepsilon} \frac{\varepsilon}{k} G_k - C_{2\varepsilon} \rho \frac{\varepsilon^2}{k} - R_\varepsilon \tag{5}$$

In Eqs. (4-5), a_k and a_ε are the inverse effective Brandt numbers for the k and ε equations, respectively; G_k is expressed by the mean flow velocity gradient arising from the turbulent kinetic energy; $C_{1\varepsilon}$ and $C_{2\varepsilon}$ are constants with values of 1.42 and 1.68, respectively; and μ_{eff} is the validity viscosity coefficient defined as follows:

$$\mu_{eff} = \mu_t = \rho C_\mu \frac{k^2}{\varepsilon} \tag{6}$$

C_μ is a constant equal to 0.0845.

The additional term R_ε is the major difference between the RNG k–ε model and the standard k–ε, which significantly improves the accuracy for a high strain rate and large degree of crook flows. This term is given as

$$R_\varepsilon = \frac{C_\mu \rho \eta^3 (1 - \eta / \eta_0)}{1 + \beta \eta^3} \frac{\varepsilon^2}{k} \tag{7}$$

Here, $\eta=Sk/\varepsilon$ and η_0 and β are constants equal to 4.380 and 0.012, respectively.

2.3 Boundary conditions

The boundaries of the computational domain are categorized as inlet, outlet, and wall. The mean velocity is specified at the inlet where the turbulent kinetic energy and specific dissipation rate are calculated according to the computational formulas of the turbulence parameter. The flow at the outlet is not defined and is allowed to change as the hydrodynamic pressure at all the boundaries is calculated from inside the domain. A stationary no-slip boundary condition is prescribed on the flume and artificial reef, and the undisturbed free surface is treated as a "moving wall", which has zero shear force and the same speed as the incoming fluid. For coarse meshes, boundary layers cannot be discretized in sufficient detail. The standard wall functions based on the proposal of Launder and Spalding was used to bridge the solution variables in the cell next to the wall to the values at the wall.

2.4 Meshing

The computational grids are shown in Fig. 1. Meshes for modeling flow fields of artificial reefs using FLUENT were created in GAMBIT. In this work, all grids used were unstructured, and the TGrid meshing scheme was used to generate unstructured tetrahedral grids. Grids around the artificial reef were generated densely, and other domains were generated sparsely to reduce unnecessary calculations.

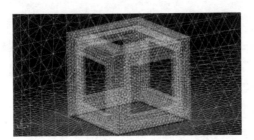

Fig. 1. Computational grids

2.5 Finite volume discretization

The equations were solved using a FVM, where the equations were integrated over a control volume. The control volume integrals were discretized into sums, which yielded discrete equations. The discrete equations were applied to each control volume. The values of the variables in the equations were solved by upwind discretization schemes. The basic concept of upwind schemes is that the face value of a given variable is defined from the cell center value in the cell upstream of the face. Upstream is defined relative to the fluid velocity normal to the face. The second-order upwind schemes were employed for the turbulent kinetic energy and turbulent energy dissipation rate here. In FLUENT6.3, all computational work was performed with the 3D pressure-based solver in a first-order implicit unsteady formulation. Gradients were estimated by the Green-Gauss cell-based method. The

SIMPLEC algorithm was employed for pressure-velocity coupling. High-resolution discretization schemes were used for the discretization of mass and momentum equations. The equations were linearized using an implicit solution and iterated to achieve a converged solution. Convergence was assumed for each time step when all residuals fall below 10^{-3}, and a maximum of 100 iterations per time step was considered if the residuals failed to pass these thresholds.

3. Experimental methods

In recent decades, PIV has become a powerful technique for measuring instantaneous velocity fields. Highly reflective particles are thrown into the flume, and a light sheet produced by a laser is projected into the flow field of interest. Then, the light reflected by these particles is captured by a high-speed camera. The calculation of the speed of the particles in the raw particle images is based on the cross-correlation, and techniques such as the erroneous vector correction are also employed to reduce the computing error. The PIV systems are described in detail below.

The structure and arrangement of a single hollow cube artificial reef model are shown in Fig. 2. In this study, the reef block model was scaled by a factor of 1:20 to satisfy the physical constraints of the flume measuring area and avoid inducing an undesirable channel wall effect. The hollow cube artificial reef model was composed of five Plexiglas faces, with a width of 7.5 cm and 4.5 cm square openings. The artificial reef model was arranged at 90^0 and 45^0 angles at the bottom of the flume. All of the test sections are based on the axial plane of the artificial reefs.

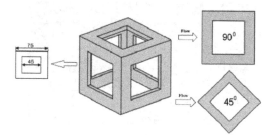

Fig. 2. Structure and arrangements of a single hollow cube artificial reef model

As shown in Fig. 3, a three-dimensional coordinate system (x-y-z) was set up to build the PIV experiment with the origin at the center of the artificial reef. The internal dimensions of the flume were 22.00×0.45×0.60 m (length ×width ×height), and the water depth of test (H) was 0.4 m. The maximum test section was 45 cm wide, 60 cm deep and 100 cm long. The sides and bottom of the test area in the center of the water channel were composed of glass to facilitate PIV measurement of the flow fields at various positions around the artificial reef model. A centrifugal pump was fixed on the left of the flume and affords flow velocities of U_1=6.7, U_2=11.0 and U_3=18.0 cm/s. The experimental flow velocities were chosen according to the characteristic flow rate of the practical sea area, and the corresponding physical values were 0.3 m/s, 0.5 m/s and 0.8 m/s. The inlet flow velocity was measured using acoustic Doppler current velocimetry.

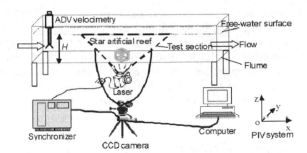

Fig. 3. Schematic diagram of the experimental apparatus and the PIV system

The basic optical devices of the PIV system were designed according to the experimental configuration specified by TSI Inc (USA). The light sheets were generated by an Nd: YAG laser capable of producing 3-5 ns, 120 mJ pulses at a repetition frequency of 15 Hz. The digital images were captured using a high-resolution CCD camera with two million pixels. The maximum frame rate of the camera was 32 f/s. A CCD camera (Power View 4MP) coupled to PC image acquisition software was used to acquire images. The operation of the laser and the camera was synchronized by a digital delay pulse generator. Selecting reflective polyvinyl chloride powder with a mean diameter of 10 μm and density of 1050 kg/m^3 was added to the water as a trace particle.

4. Flow field characteristics of a single hollow cube artificial reef

It is known that the structure of flow field within and around artificial reefs plays an important role in their ecological effects. The flow field effect is one of the most important ecological effects, therefore making artificial reefs with configurations specifically designed to induce suitable flow field structures has been of great interest to marine ecologists and engineers in recent decades. The upwelling and back vortex flow is the major flow field character, which is an important hydrodynamic characteristic of artificial reefs. Some nomenclatures are listed for analyzing and discussing these flow field characteristics qualitatively and quantitatively. The upwelling field is defined where the flow velocity of

---

Nomenclature

U_{in}: inlet current velocity (cm/s)

H: the depth of water (cm)

H_{up}: the height of upwelling field (cm)

S_e: the area of back vortex flow (cm^2)

V_{max}: the maximum upwelling current velocity (cm/s)

S: the incident flow area of artificial reef model (cm^2)

S_{eA}: unit reef effect for the area of back vortex flow field

L_{eA}: unit reef effect for the length of back vortex flow field

h: the height of artificial reef model (cm)

r: the ratio of reef height to water-depth

S_{up}: the area of upwelling field (cm^2)

L_e: the length of back vortex flow (cm)

V_a: the average upwelling current velocity (cm/s)

S_{upA}: unit reef effect for the area of upwelling

H_{upA}: unit reef effect for the height of upwelling field

L_{rp}: the length of water current reattachment point from the end of reef model (cm)

the axis z (height) orientation is equal to or greater than ten percent of inlet flow velocity. A clockwise rotation vortex is generated at the rear of the artificial reef and is called the back vortex flow field. Compared with the inlet flow velocity, the flow velocity in the back vortex flow field is slower. Therefore, the back vortex flow field is also called the slow flow region. The water current reattachment point is an important characteristic quantity and is located downstream of the artificial reef where the flow velocity is on the verge of zero. All of the arguments for the upwelling current and back vortex flow are on the vertical two-dimensional central plane.

4.1 Flow field of a single hollow cube artificial reef

4.1.1 A single hollow cube artificial reef at a 90 degree angle

Figs. 4(a) and 4(b) show the experimental and numerical results for the flow field velocity vector diagram of a single hollow cube artificial reef with a 90 degree angle, respectively. In front of the artificial reef, the flow velocities slow down due to the reef resistance causes. Several narrow regions with different values of flow velocities spread in the upstream field, and the flow velocities increase gradually with height. Within the reef, the upper and bottom regions of the flow field have a low velocity, while the regions in the opening have a high velocity, which is approximately the incoming flow velocity. The reason is that the flow into the reef is obstructed by the two vertical members on both sides of the cube side face but can enter freely through the opening located in the center of the face. Figs. 4(a) and 4(b) also show that a low velocity flow field is formed in the area downstream of the reef. Experimental and simulation results show a similar velocity distribution around the artificial reef. The overall velocity vector diagrams of the two figures are in good agreement.

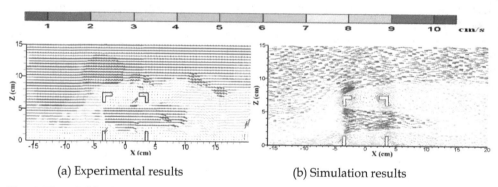

(a) Experimental results   (b) Simulation results

Fig. 4. Flow field velocity vector diagram of a hollow cube artificial reef with a 90 degree angle

Fig. 5 compares the scale and intensity of upwelling and back vortex flow field between the experimental and simulation results at a 90 degree angle. The height (H_{up}) and the area (S_{up}) of upwelling field by simulation are close to the experimental data. The computational average upwelling current velocities (V_a) are slightly higher than the experimental results. A major gap exists in the length of the back vortex flow (L_e), especially with a lower inlet flow velocity. In the field of the back vortex flow, the flow velocity is very slow and is on the verge of zero at the water current reattachment point, such that the performance of the

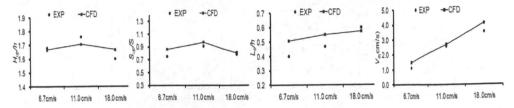

Fig. 5. H_{up}/h, S_{up}/S, L_r/h and V_a of experimental and simulation results at a 90 degree angle

tracing particles is weakened and causes a larger error. The mean error of upwelling and back vortex flow between the experiment and simulation is lower than 11%, and both of the results have an identical variation trend with the increase of the incoming flow velocity.

4.1.2 A single hollow cube artificial reef at a 45 degree angle

Figs. 6(a) and 6(b) present the experimental and numerical results obtained for the velocity vector diagrams of a hollow cube artificial reef with 45° angle. The flow structures observed at this angle are not the same as those for a 90° angle, especially inside the artificial reef. At this angle, the water flow is obstructed by the side face of the artificial reef, and no fluid can directly enter the reef in the middle. In the upstream, the slow flow velocity field is mainly distributed at the bottom or close to the reef side face. The flow velocity, both in the experiment and simulation, in most regions within the reef, is less than 2.0 cm/s, and the maximum velocity is only half of the inlet flow velocity. There is a large difference for the flow field structure in the rear of the artificial reef between the two arrangements. A large and complicated vortex is formed immediately downstream of the reef. The low flow velocity conditions in this region indicate the presence of a desirable shading effect, i.e., fish can rest and seek shelter in this suitable environment. For Fig. 6 (a) and (b), the numerical simulation predicts a flow field that is comparatively similar with the experimental results.

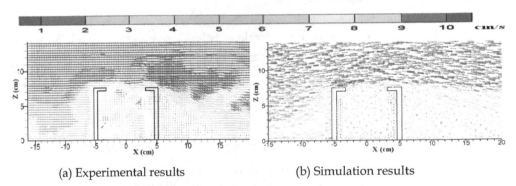

(a) Experimental results   (b) Simulation results

Fig. 6. Flow field velocity vector diagram of a hollow cube artificial reef with a 45 degree angle

Fig. 7 shows the comparisons of the scale and intensity of the upwelling and back vortex flow field that are derived from the experiment and numerical simulation at a 45 degree angle. A good agreement was found between the two results. Nonetheless, there is a large

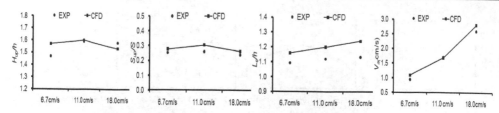

Fig. 7. H_{up}/h, S_{up}/S, L_e/h and V_a of experimental and simulation results at a 45 degree angle

difference in the comparison of the height of upwelling field when the inlet flow velocity is 6.7 cm/s, which may be caused by the lack of particle concentration in the experiment. The mean error between the experiment and simulation was within 8%. In this arrangement, the resistance to the water flow is intensified, and a large pressure gradient is generated around the reef so that a large-scale back vortex flow is produced.

4.1.3 Brief conclusion

In this small chapter, the flow field around a single hollow cube artificial reef is analyzed qualitatively and quantitatively by physical experiment and numerical simulation. The flow field within and in back of the artificial reef with a 45 degree angle has a lower flow velocity than a 90 degree angle. The scale and intensity of the upwelling field of artificial reef with a 90 degree angle are larger than a 45 degree angle. However, a biggish area of slow flow field is obtained at a 45 degree angle. The results of numerical simulation show good agreements with experimental measurements which provide powerful support for the next numerical analysis with regard to the effect of reef height on the flow field of a single artificial reef and the effect of spacing on the parallel and vertical binary combined artificial reefs.

4.2 Effect of reef height on the upwelling field and back vortex flow field

4.2.1 Distributions of the upwelling field

Fig. 8 shows that, with the same inlet velocity, the variation of reef height have nearly no effect on the general structure of the upwelling field. The upwelling field is mainly distributed at the top left corner of the artificial reef and has a fan shaped. The values of the upwelling flow velocity gradually decrease from inside to outside. Generally speaking, the height, breadth and area of the upwelling field always increase with the height of the artificial reef, and these rules are the same as those of a solid cube artificial reef (Pan et al., 2005).

4.2.2 The effect of the reef height on the intensity of the upwelling field

The effect of the reef height on the intensity of the upwelling field was measured by using V_{max} and V_a. The normalized maximum and average upwelling flow velocities are defined as the ratio of the flow velocity to the inlet current velocity, i.e., V_{max}/U_{in} and V_a/U_{in}.

According to Fig. 9, the normalized maximum upwelling flow velocity first increases and then decreases with increasing reef height and peaks at 8.0~9.0, where the value of V_{max} is close to inlet flow velocity. The maximum upwelling flow velocity remains stable when the reef height is less than 7.0; afterwards, it increases rapidly until 9.0. The normalized average

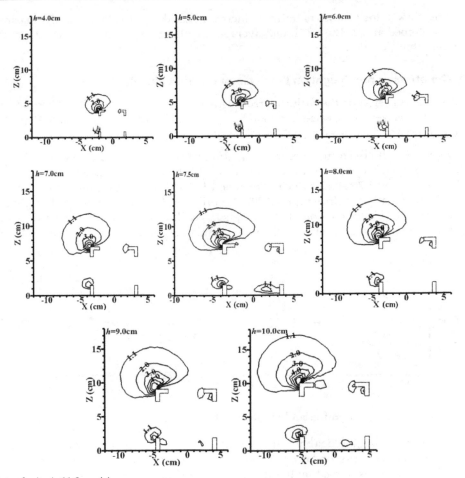

(Inlet velocity is 11.0 cm/s)

Fig. 8. Distribution of the upwelling field with different heights of hollow cube artificial reefs

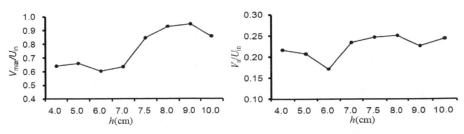

Fig. 9. Normalized maximum and average upwelling velocities with different heights of artificial reefs

upwelling flow velocity does not always increase with the reef height, and the maximum value is obtained at 8.0. The maximum average upwelling flow velocity is nearly a quarter of the inlet flow velocity.

4.2.3 The effect of reef height on the scale of the upwelling field

To find the best size of the artificial reef, a new conception unit artificial reef effect is introduced. It is weighed by length and area, and the unit artificial reef effect relative to the height and area of the upwelling field is represented by $H_{upA}=H_{up}/h$ and $S_{upA}=S_{up}/S$.

As shown in Fig. 10, the result of regression found that the height of the upwelling field and the reef height have a positive linear relationship. The best fit equation between H_{up} (axis y) and h (axis x) is y=1.76x-0.62, which is obtained by the linear regression analysis method, and the R² (correlation coefficient) is 0.9997. According to Fig. 10, when the reef height is 7.5 cm and 10.0 cm, H_{upA} (the ration of H_{up} to h) is large. Thus, the preferable unit artificial reef effect relative to H_{up} was acquired at the reef height of 7.5 cm and 10.0 cm.

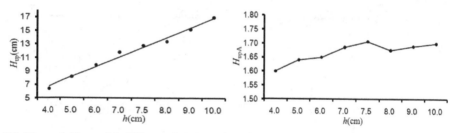

Fig. 10. H_{up} and H_{upA} with different heights of artificial reefs

The area of the upwelling field and the reef height were fitted by the nonlinear regression analysis method, i.e., a second-order polynomial regression. The relationship between S_{up} (y) and h (x) can be expressed by the equation y=0.86x²-2.51x+3.87 and R² = 0.9979. Fig. 11 shows that the curves of S_{upA} are the same as those for H_{upA}, and both of them acquire a large value at the reef heights of 7.5 cm and 10.0 cm. Thus, a preferable unit artificial reef effect relative to S_{up} is acquired at the reef heights of 7.5 cm and 10.0 cm.

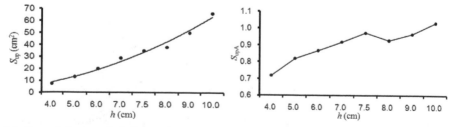

Fig. 11. S_{up} and S_{upA} with different heights of artificial reefs

4.2.4 The effect of reef height on the back vortex flow field

The length of the back vortex flow is almost equal to the distance of the water flow reattachment point from the original point, which is a significant characteristic quantity for back vortex flow. Consequently, L_{rp} and S_e are used to measure the scale of the back vortex flow field. The unit artificial reef effect relative to the length and area of the back vortex flow field is represented by L_{eA} and S_{eA}.

In Fig. 12, the length from the water current reattachment point to the artificial reef and the area of back vortex flow increase continually with increasing reef height. L_{rp} with h (in x axis) presents an exponential growth relationship, and the relevant relational equation is $y=1.09e^{x/5.46}+0.20$. A good correlation coefficient of 0.9976 is acquired. The quadratic polynomial fitted equation for the area of the back vortex flow (y) and reef height (x) is expressed by $y=0.21x^2-1.11x+3.03$, and the homologous correlation coefficient is 0.9973.

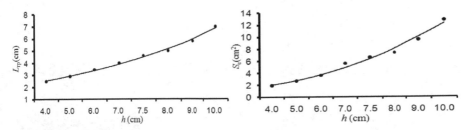

Fig. 12. L_{rp} and S_e with different heights of artificial reefs

As shown in Fig. 13, the variation curves of L_{eA} and S_{eA} with reef height are similar to some degree. The values of L_{eA} and S_{eA} are the minimum at the reef height of 6.0 cm; after that point, there is a relatively fast increase, and a steady value is maintained until 9.0. There are three large values appearing at the heights of 4.0, 7.5 and 10.0, which can be used for the artificial reef designers as a reference.

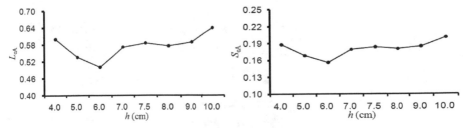

Fig. 13. L_{eA} and S_{eA} with different heights of artificial reefs

4.2.5 Brief conclusion

Considering the economical problems and the stability of artificial reef in a certain depth of water, the height of artificial reef cannot be raised excessively. According to the results calculated in the numerical simulation, when the reef height is 7.5cm, the efficiency of the reef unit performs best.

4.3 Flow field velocity vector diagram in the horizontal cross-section

The velocity vector diagrams in the horizontal cross-section that is located at the height of $1/2h$ are listed in Fig. 14. Similar structures of the flow field are obtained with different heights of artificial reefs. Within the reef, the flow velocity is close to the inlet velocity. In the area on either side of the artificial reef, the horizontal flow velocity is larger than the inlet flow velocity, and the maximum values are 11.5, 11.9, 11.9, 12.3, 12.9 and 12.9 cm/s, respectively, with increasing reef height. It is apparent that a higher artificial reef will have a wider area in which the horizontal flow velocity is greater than the inlet velocity. The slow flow field is distributed in the front and the back of the reef, where the structures of flow fields vary with the reef height.

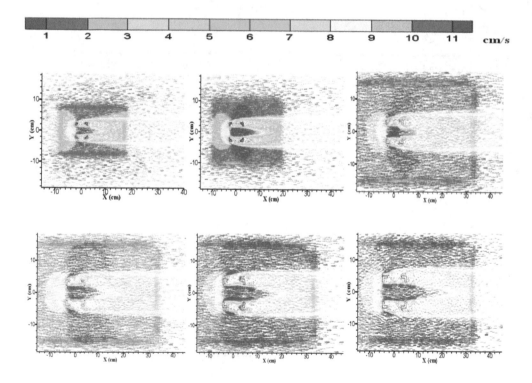

Fig. 14. Velocity vector diagrams in horizontal cross-section with different heights of artificial reefs (the inlet velocity is 11.0 cm/s)

The horizontal flow velocities along the line of y=0 at the horizontal cross-sections of z=1/2h are shown in Fig. 15. Similar distributions of flow velocities are found within and around artificial reefs at different heights. In front of the artificial reef, the horizontal flow velocities are approximate and decrease gradually when approaching the artificial reefs. Within artificial reefs, the flow velocities are larger than the inlet flow and peak at the center of the reefs. In addition, with the increase of the reef height, the maximum horizontal flow velocity within the reef is also on the increase. At the rear part of the artificial reefs, the flow velocities show a long-distance attenuation until the locations are sufficiently far away from

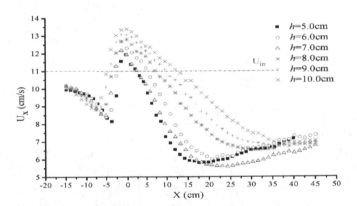

Fig. 15. Distribution of the horizontal flow velocities along the line of Y=0 at the horizontal cross-section of Z=1/2h (h=5.0, 6.0, 7.0, 8.0, 9.0, 10.0 cm)

the reefs, which are the slow flow regions as mentioned earlier. The smaller the artificial reef is, the lower the horizontal flow velocity is at the rear of artificial reefs. However, there is a special phenomenon where the lowest velocity occurs when the reef height is 7.0 cm. In other words, a preferable slow flow effect is acquired at the reef height of 7.0 cm.

5. Flow field characteristics of two hollow cube artificial reefs

The arrangements of two models of artificial reef are shown in Fig. 16. The spacing of two parallel combined models is 0.5h, 1.0h, 2.0h and 3.0h. The spacing of two vertical combined models is 1.0h, 2.0h, 4.0h, 6.0h, 8.0h, 10.0h and 12.0h (h=7.5 cm).

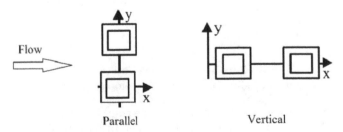

Fig. 16. The arrangements of two hollow cube artificial reefs

5.1 Two parallel hollow cube artificial reefs

Fig. 17 presents the numerical results for the upwelling field of single and two parallel artificial reefs with different spacings. Compared to a single artificial reef, the heights of the upwelling field increase with spacing at first and then decrease until steady. When the spacing is 0.5h, the height of the upwelling field reaches the maximum value of 13.8 cm, and the corresponding area is 49.0 cm². When the spacing is equal to or larger than 2.0h, the height of upwelling field has a steady value of 12.5 cm, and the steady value of the area is 35.0 cm².

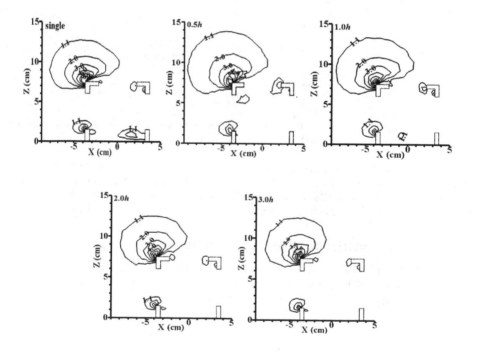

Fig. 17. Distribution of the upwelling field of a single and two parallel hollow cube artificial reefs with different spacings by simulation (the inlet velocity is 11.0 cm/s)

Fig. 18 shows a long and narrow slow flow field spread at the rear of a single artificial reef. Along the forward direction of the x-axis, the maximum horizontal flow velocity appears within the artificial reef, and after that, the flow velocity first decreases and then increases.

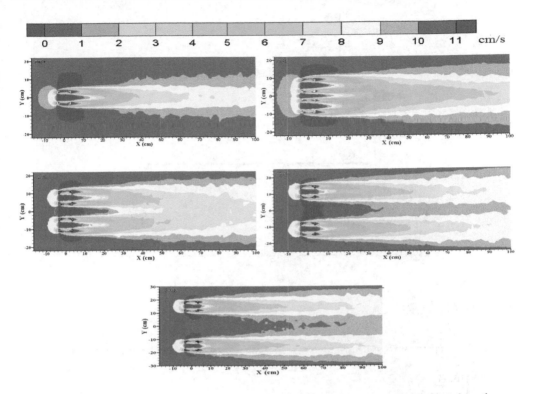

Fig. 18. Horizontal flow velocity contour plots of a single and two parallel artificial reefs with different spacings (the horizontal cross-section is z=1/2h, h=7.5 cm)

The mutual effect of two reefs is very strong, and the slow flow velocity fields in the front and the back of two artificial reefs nearly connect when the spacing is 0.5h. Thus, a larger area of slow velocity flow field is produced for a 0.5h gap distance. When the spacing increases to 1.0h, downstream of the reefs, the flow speed begins to increase, and the construction of the flow field presents a trend of separation. The slow velocity flow fields behind two artificial reefs are separated from each other by and large when the spacing is 2.0h. After that point, the structure of the flow field for each artificial reef becomes the same as the single condition, and the interaction between the two reefs is slight when the distance increases to 3.0h.

5.2 Two vertical hollow cube artificial reefs

Fig. 19 compares the upwelling field of two vertical artificial reefs with different spacings. As shown in the graphs, for different spacings, the latter artificial reef causes little impact on the upwelling field of the first artificial reef. With different spacings, the scales of the upwelling field of the first reef are the same as that of a single reef. Nevertheless, with the increase of the vertical spacing from 2.0h to 12.0h, the height of the upwelling field of the second incident flow artificial reef increases gradually. When the spacing is 12.0h, the height of the upwelling field of two artificial reefs is close, but the area gap is greater.

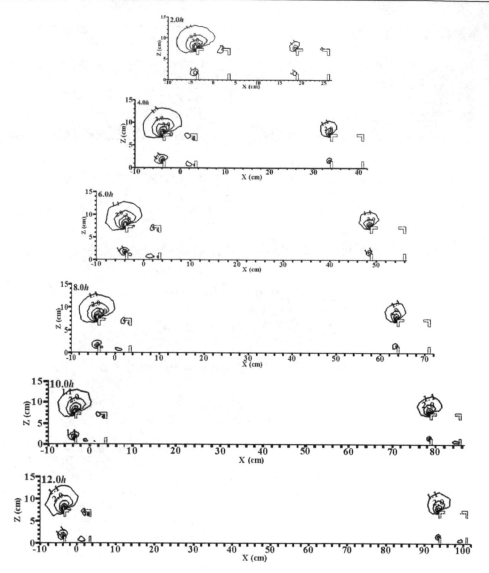

Fig. 19. Distribution of the upwelling field of a single and two vertical hollow cube artificial reefs with different spacings by simulation (the inlet velocity is 11.0 cm/s)

As shown in Fig. 20, the flow fields between two reefs are constantly changing with increasing spacing. The horizontal flow field structure of two artificial reefs with 1.0h spacing is similar to that of a single reef. With the spacing of two reefs added to 2.0h, the construction of the flow field presents a trend of separation. When the spacing is 4.0h, the flow velocity in most areas between the two reefs is slow after that the flow velocity increase gradually with the spacing. The interactions of two artificial reefs are first strengthened and then weakened with increasing distance.

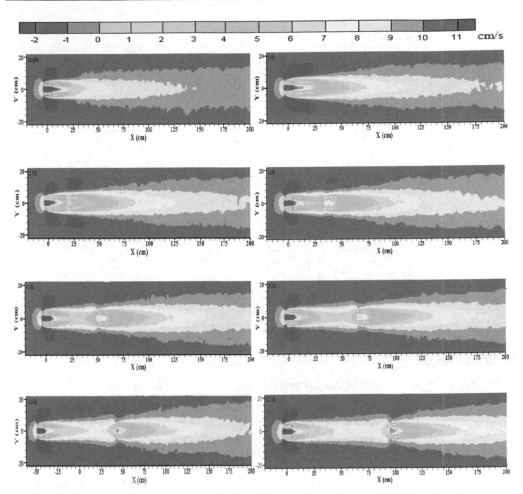

Fig. 20. Horizontal flow velocity contour plots of a single and two artificial reefs with different spacings (Z=0.5h)

Fig. 21 shows that the flow velocity within and in the front of the first artificial reef with different spacings is approximately equal to that of a single reef; however, the flow velocities in the rear of two reefs are less than that of the single condition. The same distribution regularity of the flow velocity between two artificial reefs is obtained for different spacings. The flow velocity decreases first and then increases suddenly in the front field of the second artificial reef. The same distribution rules for flow velocities also exist in the rear of the two reefs for different spacings. When the spacing is 4.0h, the minimum flow velocity in the area between the two artificial reefs has a minimum value; after that point, it increases with spacing. It was also found that the minimum velocity in the back of the two reefs is acquired at the spacing of 2.0h, and a large gap was found compared with other conditions. After the key point in the back field of two reefs where the flow velocity is the minimum, a larger spacing leads to a lower flow velocity.

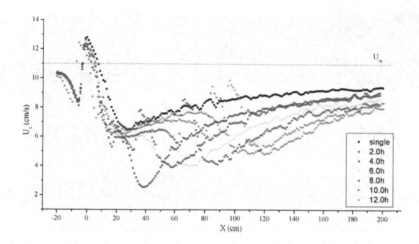

Fig. 21. Distribution of the horizontal flow velocities of two vertical artificial reefs along the line of Y=0 at the horizontal cross-section of Z=1/2*h* (*h*=7.5 cm)

6. Conclusions

Some main conclusions can be drawn as follows:

1. The flow field within and around a single hollow cube artificial reef was investigated by physical experiments and numerical simulations. The upwelling and back vortex flow fields that arose from artificial reefs were analyzed by using qualitative and quantitative analysis methods. According to the comparisons, there is a good agreement between the numerical and experimental results.

2. Based on the experimental verification, a detailed numerical study was conducted to investigate the effect of reef height on the structure of the flow field. The height and area of upwelling current region increased linearly and nonlinearly with increasing reef height, respectively. When the value of r (the reef height to water-depth) was 0.2, the normalized maximum and average upwelling flow velocities reached their maximum values. An exponential growth relationship existed between the length of the back vortex flow and the reef height. The flow field on either side of artificial reef was also discussed. A preferable unit artificial reef effect was achieved when the ratio of the reef height to water-depth was 0.2.

3. In the meantime, the influence of the arrangement (parallel and vertical) and spacing on the flow field of the two artificial reefs were discussed using the results of the numerical method. In the parallel combination, the scale of the upwelling field reached the maximum value, and a larger area of slow velocity flow field was obtained. The interaction between the two reefs was slight after the spacing became larger than 2.0*h*. In the vertical combination, the latter artificial reef had a weaker effect on the distribution of flow velocities within and in the front of the first artificial reef. When the

spacing was $4.0h$, the flow velocity in the area between the two reefs obtained the minimum value. Behind the two reefs, the lowest flow velocity was achieved at a spacing of $2.0h$, and, at a larger spacing, there was a lower flow velocity.

7. Acknowledgments

This work was financially supported by the National Natural Science Foundation (NSFC) Project No.50809014 and 50921001, the National 863 High Technology Project No.2006AA100301, the Specialized Research Fund for the Doctoral Program of Higher Education No.200801411094 and Special Fund for Agro-scientific Research in the Public Interest Project NO.201003068.

8. References

Baine, M. (2001). Ocean & Coastal Management. *Artificial reefs: a review of their design, application, management and performance,* Vol.44, No.3-4, pp. 241-259

Einbinder, S.; Perelberg, A. & Ben-Shaprut, O, et al. (2006). MARINE ENVIRONMENTAL RESEARCH. *Effects of artificial reefs on fish grazing in their vicinity: Evidence from algae presentation experiments,* Vol.61, No.1, pp. 110-119

Hixon, M.A. & Beets, J.P. (1989). BULLETIN OF MARINE SCIENCE. *SHELTER CHARACTERISTICS AND CARIBBEAN FISH ASSEMBLAGES: EXPERIMENTS WITH ARTIFICIAL REEFS,* Vol.44, No.2, pp. 666-680

Jensen, A. (June 1998). *European Artificial Reef Research Network (EARRN): Final Report and Recommendations.* Published by the University of Southampton.

Leitao, F.; Santos, MN. & Erzini, K, et al. (2008). MARINE BIOLOGY. *The effect of predation on artificial reef juvenile demersal fish species,* Vol.153, No.6, pp. 1233-1244

Lin, J. & Zhang, S.Y. (2006). Marine Fisheries. *Research advances on physical stability and ecological effects of artificial reef.* Vol.28, No.3, pp. 257-262

Liu, H.S.; Ma, X. & Zhang, S.Y., et al. (2009). *Validation and comparison between wind tunnel experiments and numerical simulation of flow field around artificial reefs* Vol.16, No.3, pp. 365-371

Pan, L.Z.; Lin, J. & Zhang, S.Y. (2005). Journal of Shanghai Fisheries University. *A numerical experiment of the effects of artificial reef on vertical2-dimensional steady flow field.* Vol.14, No.4, pp. 406-412

Seaman, W. (2007). Hydrobiologia. *Artificial habitats and the restoration of degraded marine ecosystems and fisheries,* Vol.580, No.1, pp. 143-155

Su, D.T.; Liu, T.L. & Ou, C.H. (2007). *A Comparison of the PIV Measurements and Numerical Predictions of the Flow Field Patterns within an Artificial Reef.* In: 17th international offshore and polar engineering conference, Lisbon, Vol. 1-4. pp. 2239-2245.

Walker, B.K.; Henderson, B. & Spieler, R.E. (April 2002). Aquatic Living Resources. *Fish assemblages associated with artificial reefs of concrete aggregates or quarry stone offshore Miami Beach, Florida, USA,* Vol.15, No.2, pp. 95-105

Williams, T.W. (August 2006). *Sinking Poor Decision Making With Best Practices: A Case Study of Artificial Reef Decision-Making in the Florida Keys,* Virginia Commonwealth University, America.

Zhang, H.H. & Sun L. (2001). Resources Science. *On reproduction increase of the sea aquatic resources by artificial fish-reef engineering*. Vol.23, No.5, pp. 6-10

Part 2

Water Management

6

Aquaculture Water Quality
for Small-Scale Producers

Oscar Alatorre-Jácome, Fernando García-Trejo,
Enrique Rico-García and Genaro M. Soto-Zarazúa
C.A. de Ingeniería de Biosistemas, Campus Amazcala,
Facultad de Ingeniería, Universidad Autónoma de Querétaro,
Amazcala, Mpio, El Marqués,C.P.76260, Querétaro, Qro
México

1. Introduction

Today we known that solar energy can be stored as edible biomass by photosynthesis, but 10,000 years ago this assumption was unthinkable. However, the Neolithic man could developed agriculture, based on plant germination observations. Nowadays, the systems used for food production are practically the same (Koning et al., 2008). But for neolitican man, the empirical observation of germination processes and biological cycles of certain species ended in the discovery of agriculture.

Even our agriculture is still the same, the climate scenario is changing. Our current carbon-based economy has caused massive greenhouse emissions, and consequences are global warming and the depletion of the ozone layer. The biological, economical and social impact on the environment is significant. Water scarcity and natural catastrophes (tsunamis, earthquakes, floods and erosion) damage cultivable land, reducing local food availability (Franck et al., 2011, IPCC, 2007).

Clearly, it is necessary the innovation of new food production systems. Aquatic food production is an alternative to land-based, 2-dimensional systems (Alagaraja, 2007). Due to physical differences between air and water (thermal conductivity, heat capacity, density, etc), most aquatic organisms do not use energy in thermal regulation. The consequence is a more efficient food-tissue conversion ratio compared with land-based system yield (FAO, 2010).

There are limitations, of course. Because overfishing causes fish stock depletion, aquaculture has become an alternative to provide fish products (Naylor, 2009; Brander, 2007; Pillay et al., 2003; Naylor, 2000).

The notorious growth in aquaculture has been possible by the application of science. Aquacultural and fisheries science produce new and replicable knowledge based on scientific method. The research is usually done in farms or in a laboratory. The mathematical and analytical methods used by researches are generally robust and provided a verifiable knowledge. Nevertheless, there is a gap between the scientific results and its practical use. Lukefar (1999) pointed out that the examples taught on classroom lectures are usually based under realistic, idealized situations. But there is a necessity to find practical solutions applicable for commercial aquaculture.

The problem of the academic extension is an important issue in aquaculture if we consider that small-scale aquaculture provides almost half of the worldwide inland fish production (FAO, 2010). Kawarazuka (2010) commented "aquaculture and small-scale fisheries can improve the food intake". Mohanty et al. (2010) mentioned that at least 90 % of the people involved in aquacultural practices works in small fisheries.

Facing the problem of the continuous rising world population, it seems to be clear that the good use of all natural resources (including small freshwater bodies), must be done. So there is an important, unattended practical knowledge field to be considered. Technologies for small-scale fish producers are ready available (Van Gorder, 2003), and the current knowledge on aquatic science can be helpful for the producer with the goal to optimize its own resources (time, land, water and energy).

The objective of the present chapter is to present selected topics on water quality management, especially if they can be carried out by low-cost technologies in small systems. Due to the close relationship among water quality and yields, we hope it can be helpful as a practical guide for fish producers based on scientific principles.

2. Overview of water quality in aquaculture

Water is the physical environment where fish develop, growth and reproduce. The dynamic of the mass and energy involved on an aquaculture system is complex, because bacteria, algae and fish growth together in the pond (Wheaton, 1982). The main energy input comes from sunlight, and nutrients of the system are commonly provided by pelletized fish feed. So the transformation of the elements carried out by autotrophic and heterotrophic organisms change the physical/chemical/biological composition of the water (Hargreaves, 1998).

Temperature and pH are fundamental for the aquatic living organisms, due to the intimate relationship between them and the velocity of its biochemical processes. Oxygen (O_2) and carbon dioxide (CO_2) are important molecules because they are involved in photosynthesis and respiration processes. Nutrients like nitrogen are essential on biological metabolism: when is ingested as protein or amino acid, it can be incorporated by the organism as functional proteins or in structural tissues. But when is excreted by the fish as ammonium (NH_3), in certain circumstances it can be toxic and even lethal in high doses. Another nutrients, like phosphorus, potassium and calcium are also important, and they lack are usually diagnosed by deficiency.

Besides the physical and chemical factors, the biotic component also can change the water composition. For example, algae can consume of produce oxygen and carbon dioxide, depending on light presence or absence. The nitrogen, phosphorus and potassium can be used and assimilated by unicellular organism, incorporating them as biomolecules. Moreover, the nitrogen can be also used to produce energy on chemoautotrophic biochemical cycles (Hargreaves, 1998).

So, the biological impacts of water quality over the cultivated species could be analyzed under physical, chemical or biological perspectives. For practical purposes, a quick overview of the basic principles is given below. The parameters mentioned have been divided into two main categories: Critical and important parameters.

2.1 Critical water quality parameters

The critical parameters are very important in the aquacultural system. They are temperature, pH, dissolved oxygen (DO) and ammonia. They must be measure daily, or in the case of intensive systems, all day long. They do influence the physical properties and chemical composition of the water, and thereafter its correct management can improve the overall fish performance (health and growth). In the other hand, if they are not properly attended, the consequences can be serious, varying from poor growth rates, stress, and death.

2.1.1 Temperature

Temperature is probably the most important physical variable on aquatic ecology. It affects directly the metabolism of all living organisms. As a consequence, temperature set the growth, development and reproduction rates in biological species in biological species. This fact is very useful in aquaculture: because fish do not expend energy on corporal temperature regulation, they can assimilate almost the food nutrients into muscular tissue (Soto-Zarazúa et al., 2011). As results of an adequate temperature condition, the biomass production and final yield of the fish farm can increase.

It is important to remark the influence of temperature in fish respiration rate. A rise of temperature causes more oxygen consumption in bacteria, algae and fish (Boyd, 1998). Because respiration implies carbon dioxide release and energy consumption, the gas balance can be dangerous for fish. If there are enough inorganic nutrients in the water, the algae biomass can increase to considerable levels. Even some algae species can double its biomass in only 3.5 hours (Brennan & Owende, 2010). Then, the elevated rates of nutrient assimilation will produce significant impacts on the water quality.

In other hand, a higher temperature produces higher metabolic rates. The increment in fish metabolism enhances the protein breakdown. As a consequence, the release of NH_3 by fish will be high, too. The resulting combination of high temperatures with NH_3 high concentration is very a toxic environment (Eshchar et al., 2006).

Finally, if the exposure to high temperatures is very long, the structure of the proteins begins to break, causing fish death.

2.1.2 pH

The pH is a measure of acidity and basicity inside an aqueous solution. It indicates the concentration of hydrogen ions on water. When pH is below 7 the solution is considered acid, and when is above 7 is named basic or alkaline. Distilled water has a pH of 7. pH can be defined as the negative logarithm of the molar concentration of hydrogen ions on water. In mathematical notation, pH is described as

$$pH = \log\frac{1}{[H^+]} = -\log[H^+]$$

Where [H+] is the concentration of hydrogen ions. Fish and other vertebrates have a pH blood value near to 7.4. The contact between environmental water and fish blood is only separated by one or two cells of the gills. An ideal pH for an aquacultural system must be near to 7. The lethal limits are below 5 and above 10, for most of the fish species.

There is an important relation between fish respiration and pH. In the gills the gaseous interchange of O_2 instead CO_2 occurs. This interchange can be difficult if pH is not optimum. The effects are called Bohr and root. So even if we have enough oxygen in our system, if pH is not adequate fish could not breathe (Wurts & Durborow, 1992).

pH, like temperature, is always changing. For example: In afternoon, the oxygen concentration decrease and the phytoplankton photosynthesis stops by the absence of sunlight. The concentrations of O_2 and CO_2 began to change, and the pH can vary due its intimate relationship with CO_2 equilibrium (Wurts & Durborow, 1992). Another important factor to take into account is that cinematic of certain types of bacteria can change in low pH, so the mass and energy transformations of the pool carried out by unicellular organism can influence the environment (Ebeling et al., 2006).

2.1.3 Dissolved oxygen

According to Rumei et al. (2003), dissolved oxygen is the most important manageable variable in aquaculture. The oxygen is necessary to glucose breakdown and energy release inside fish cells. However, its diffusivity and availability on water is mediated by temperature, elevation and salinity (Boyd, 1998).

Low concentrations of oxygen can produce negative impacts on fish health, like poor growth performance, low feeding rate, and increase risk on potential diseases or even fish death. These impacts are specific for every fish species. The particular oxygen requirements depend principally on the fish biology. For example, trout needs a high quantity of oxygen (about 7 ppm), but catfish (bottom, detritivore fish) can survive with only 0.5 ppm of oxygen (Akinwole & Faturoti, 2007).

The presence of dissolved oxygen on aquaculture water ponds depends on physical factors (salinity, temperature and altitude), algae photosynthesis or artificial supplying (Boyd, 1998). Oxygen consumption depends on the carry capacity of the systems. The fish biomass is an important factor to be considered in a system, because the food supplied to the fish plays an important role in water ponds biogeochemistry (Timmons, 2002; Hargreaves, 1998).

2.1.4 Ammonia

The ammonia is a nitrogen compound excreted by fish through gills and faeces. So the amount of ammonia is in direct relation with the amount of feed input on the pond. Ammonia can also be produced in pond by organic material decomposition driven by bacterial activity (Durborow et al., 1997a).

The ammonia is presented in two forms: the high toxic un-ionized ammonia (NH_3), and the non-toxic ionized ammonia (NH_4^+). They are in chemical equilibrium driven by temperature and pH. The sum of NH_3 and NH_4^+ is called Total Nitrogen Ammonia (TAN). In general, the fraction of toxic ammonia increases at elevated temperature and high values of pH (Durborow et al., 1997a), as it can be seen in the following equation:

$$\frac{NH_3}{TAN} = \frac{K_{eq}}{10^{-pH} + K_{eq}}$$

Where k_{eq} is the negative logarithm of pK ($pK = -\log(K_{eq})$), and $pK = 0.09018 + 2727.9 - (T + 273.1) + (0.1552 + 0.0003142T)I$, T is temperature (°C) and I is Ionic strength (M) (Eshchar et al., 2006). For more convenient purposes, computations of ammonia toxic fractions can be made with the tables provided by Durborow et al. (1997a). Measures of TAN, pH and temperature are required.

2.2 Important water quality parameters

The dynamic of the important parameters occurs slower than the reactions implicit on critical ones, so they can be monitored with less frequency. Solids, nitrite (NO_2), nitrate (NO_3), carbon dioxide (CO_2), hardness and alkalinity can be measure once a week. Lethal effects on the fish at high concentrations of this parameters are not common, but their accumulation can affects directly and indirectly the fish growth. Also, if unfavourable conditions are presented, an enhanced risk to infections or diseases can be presented.

2.2.1 Solids

In aquaculture, the solids are placed into a special category because in most of the cases they can be controlled by good management practices. Solids in pond are presented as uneaten food, faeces, fish scales, dead bacteria and algae, dust, and dead fish (Cripps and Bergheim, 2000). The adverse consequences driven by the presence of organic solids in the system are caused mainly by bacterial processes. It implies additional oxygen consumption and carbon dioxide release, among other effects. In some anoxic environments, hydrogen sulfide (H_2S) and ammonia can be release to water by bacterial bacterial anaerobic processes.

In general, the solids can be classified in three major divisions. 1) Settleable solids, wich can be easy separated by sedimentation or decantation (for example, uneaten feed, fish scales, faeces); 2) Suspended solids, which are very fine solids without the capacity of rapid sedimentation (about 1-10 μm on diameter); and 3) dissolved solids, which are nano-scale elementary forms as molecules or atoms. (Losordo et al., 1999; Malone, 1991).

In general, the impacts of the solids in aquaculture systems are negative. For example, in the surface when the water turbidity rises, the photosynthetic activity in algae decreases. In bottom, accumulation of solid wastes causes anaerobic zones, in which undesirable bacteria can proliferate. In ponds, high clay turbidity usually causes acidity, low nutrient levels, and limited light penetration for photosynthesis (Yi et al., 2003). Dead fish are solids too, and its presence on culture water can be a factor for pathogenic propagation inside the farms (Cripps & Bergheim, 2000).

Is useful to think that solids on aquacultural environments are mainly composed by organic material. If they remain in the water, they became basics nutrients, like nitrogen and phosphorus that could be easy assimilated by microscopic organism. So the total oxygen budget will rise (Timmons et al., 2002).

2.2.2 Nitrite

The presence of nitrite inside an aquaculture pond is often caused by ammonia biological oxidation (Hargreaves, 1998). The stoichiometric reaction is the following:

$$NH_4^+ + 1.5O_2 \rightarrow NO_2^- + 2H^+ + H_2O$$

As a result, there is a free energy yield (Δ G) of – 65 kcal mole[-1] from the ammonia oxidation.

This reaction is carried out by bacteria that naturally growths in the aquacultural environment. Bacteria of the genus *Nitrosomonas* are the main responsible for nitrification in aquaculture, but there are other important genus involved in this processes (Hovaneck & DeLong, 1996; Ebeling et al., 2006).

The physiological effects of nitrite in the fish health are mainly caused by a chemical reaction on the haemoglobin. The role of this protein is the oxygen transportation all over the blood stream. In the presence of nitrite, haemoglobin becomes methaemoglobin, a non-efficient oxygen transporter (Jensen, 2003). As a result of oxygen deficit, the fish blood became brown, and a gasping behaviour can be observed. If nitrite intoxication remains unattended, massive fish dead became after a short time caused by hypoxia (Masser et al., 1999).

High nitrogen concentrations in ponds occurs more frequently in the fall and spring, when low and fluctuating temperatures cause decay rates on phytoplankton and bacteria metabolisms (Durborow et al., 1997b).

2.2.3 Nitrate

The presence of nitrate in aquaculture water is water is delivered as a waste product of organic bacterial activity in the pond. The reaction involves the presence of nitrite and oxygen. The following stoichiometric reaction shows the overall process (Wheaton, 1982).

$$NO_2^- + 1.5O_2 \rightarrow NO_3^-$$

In total, when a nitrite molecule is oxidized, a free energy yield (ΔG) of -18 kcal mole[-1] is released. In aquacultural ponds, the most representative bacteria genus that can perform the nitrite conversion to nitrate is *Nitrobacter*, but other genus of bacteria are commonly presented during the nitrification (Camargo et al., 2005).

The nitrate is commonly controlled in aquaculture systems by dilution. In intensive recirculation systems between 5 and 10% of water are removed and replaced every day. In systems with low technification, the daily water exchange usually is more than 10%.

A natural pathway to remove the nitrite in an aquacultural system is done by denitrification. The reaction is carried out by bacteria in absence of oxygen and in presence of methanol as a carbon compounds. The general reaction is done in two steps (Van Rijn et al., 2006):

$$NO_3 + \frac{5}{6}CH_3OH \rightarrow \frac{1}{2}N_2 + \frac{5}{6}CO_2 + \frac{2}{3}H_2O + OH^-$$

2.2.4 Carbon dioxide

Carbon dioxide (CO_2) in aquatic systems is very important, because its presence is required in some chemical and biological processes. For example, CO_2 interacts with water to form a natural buffer system which helps to maintain a constant pH. Is also important for biological primary production, because is necessary in algal photosynthesis to synthetize glucose. Is important to note that biological activity can also release CO_2 to the media, when fish and algae breathe. Some bacteria also need CO_2 to maintain its metabolic cycles in constant function.

Bacteria also release CO_2, both autotrophs and heterotrophs. For example: The free acid produced during nitrification reacts with bicarbonate alkalinity in water to release more CO_2 than autotrophs consume. So for every gram of TAN metabolized, 4.6 ppm of oxygen will be consumed, and 5.9 g of CO_2 will be released. In heterotrophic reactions, for every gram of O_2 consumed, 1.38 g of CO_2 is released (Summerfelt & Sharrer, 2004).

When CO_2 is dissolved in water, part of it combines to form carbonic acid (H_2CO_3). This weak acid tend to react with calcium carbonate ($CaCO_3$) to form calcium bicarbonate [$Ca(HCO_3)$], which is often dissociated on hydrogen ions and carbonate ions (Wheaton, 1982):

$$CO_2 + H_2O \leftrightarrow H_2CO_3 \leftrightarrow H^+ + HCO_3 \leftrightarrow H^+ + HCO_3^{2-}$$

These four reactions are mediated by hydrogen concentration. They proceed to the right when pH rises and they go to the left when pH decreases. Therefore, the carbon dioxide usually can be found in water under four different forms: As free gas (CO_2), as carbonic acid (H_2CO_3), as carbonate (CO_3^{2-}) and as bicarbonate (HCO_3^-) (Wheaton, 1982). Thus, the chemical importance of CO_2 relies in the fact that is a pH buffer, and it has a lot of relation with other physicochemical parameters, like hardness and alkalinity.

2.2.5 Hardness

The hardness is defined as the total concentration of calcium and magnesium ions, expressed as calcium carbonated. However, if other metallic ions are presented in the water (Al, Fe, Mg, Sr, Zn) they can be also considered in the definition (Wheaton, 1982).

The hardness is an important water quality parameter because a direct relation between water metal content and pH variations exists. When the concentration of Ca and Mg trends to be higher, the buffering capacity of the water becomes higher, too, and is more capable to smooth pH variations. In other words, hard water is more stable than soft water.

In the biological perspective, calcium is important in fish metabolism, because is used on the scale and bond formation, and to keep the adequate balance of Na and K in the blood (Wurst & Durborow, 1992). Calcium is often required in neural synapses, and in physiological ion balance. In the case of magnesium, it is used is used by photosynthetic organism because is embedded in the center of the chlorophyll molecule, and it is also required as prosthetic group in proteins (Müller-Esterl, 2008)

In the farm, hardness is very important when the organisms are cultivated for reproduction purposes in hatcheries. If invertebrates are cultivated, hardness becomes an important variable to consider, because Ca and Mg are very important in the formation of hard parts (i.e., exoskeleton or shells).

Changes in the water hardness are slowly. They take place typically in weeks or months.

However, is a good management practice to monitoring this variable constantly.

2.2.6 Alkalinity

The alkalinity is the amount of acids (H^+) that water can neutralize before to reach a given pH. It is defined as the stoichiometric sum of the bases in a solution. Common bases found in fish ponds include carbonates, bicarbonates, hydroxides, phosphates and berates. Carbonates and bicarbonates are the most common and the most important components of alkalinity. Because total alkalinity can be expressed as ppm of $CaCO_3$, is common to confuse it with hardness (Wurts & Durborow, 1992).

Like another water quality parameters, alkalinity can be affected by the biological activity of systems. Because photosynthesis of phytoplankton requires CO_2 to synthesize glucose, pH in the water increases due to inorganic carbon adsorption in water (mainly H_2CO_3 and CO_3^{2-}). During long periods of intensive photosynthesis, the release of carbonate can elevate the pH levels over 9. These effects can be observed if water has low alkalinity (20 to 50 ppm) or low bicarbonate (75 to 200 ppm). High photosynthetic activity can be presented when the sodium and potassium carbonates are dissolved in water, because they are more soluble than calcium and magnesium bicarbonates (Wurts & Durborow, 1992).

3. Low-tech water quality management

Currently, the aquacultural engineering gives a considerable number of solutions to control the variables involved in aquaculture water quality. In general, they are available commercial devices and chemical products to control water quality. However, there are alternative techniques used by scientist that considered the pool as a bioreactor. New aquaculture techniques are bio-flocs (Avnimelech, 2006), Integrated Multi-trophic Aquaculture (IMTA) (Chopin, 2003), greenwater systems (Hargreaves, 2006), Zero-Exchange water systems (Panjaitan, 2010; Olvera-Olvera et al., 2009) and aquaponics (Rakocy & Hargreaves, 1993), among others.

3.1 Temperature

The temperature control on aquaculture could be difficult, because the high specific heat capacity of water (4.18 kJ kg^{-1} °K^{-1}) implies a high amount of energy when water is heated or cooled. In addition, the water volume used on aquacultural facilities is frequently high, so the monetary and environmental cost to raise water temperature could be unaffordable (Seginer & Mozes, 2008).

There are several alternatives to increase thermal stability on aquaculture. The most effective system is the use of greenhouse to cover aquaculture ponds or tanks (Soto-Zarazúa et al, 2011; Fuller, 2007). Some farms use heating pumps, thermosolar systems, fossil fuel heaters or electric resistances, but all of them are expensive ways to heat water, both by initial and operational cost. Besides, some of them imply negative environmental impacts caused by greenhouse gas emissions (Mohanty et al., 2010).

Nevertheless, there are alternatives for the small-scale producers. The most effective strategy is a good planning on farm building. Then, the most important thing is to pick a good geographical location. The climate and the temperature of the make-up water must be adequate for the cultivated fish. An analysis of the local climatologically data, if available, can give us an idea if the selected location is proper to our intentions.

If low air temperatures are presented, covering the tanks with plastic sheets could be is useful (Van Gorder, 2003; Crab, 2009). On the other hand, if temperature rises, a packed aeration column can be built as a chiller (Wheaton, 1982). In both cases (low and high water temperatures), to add new make-up water can be useful. Is also important to remember that some fish doesn't growth on winter, like tilapia (Crab et al., 2009)

If fish tanks are used, it could be helpful to cover its walls with heat insulators. Elastomeric foams are a good choice, because they are generally designed to support extreme insulation and hard management. Fiber glass and polyotyrene foams could be used, too (Alaturre, 2010).

Another option is to consider solar heating water systems. Despite the high initial cost, they can help to raise the water temperature from 1 to 5 °C. Solar pool blankets, geodesic structures and active/passive solar collectors have been tested with considerable results (Fuller, 2007). A practical guide and cases studies is available in www.retscreen.net. A complete guide to thermosolar processes can be found on Duffie and Beckman (1992).

3.2 pH

A common action to pH control is the addition of chemical substances. When pH is low, it could be useful adding lime (Wurts and Durborow, 1992). If pH rises, small amounts of phosphoric acid or acetic acid can help to neutralize water. Some people uses sulphuric acid, but it can be dangerous if there is no previous experience on dangerous substances management. Another way to pH control is the enrichment of the biological activity in the pool (Hargreaves, 2006).

In our experience, many traditional aquaculture managers do not monitoring pH. This can be a bad choice, because as we discussed previously, a certain combination of temperature, pH and ammonium can kill entire production stocks.

In the case of aquaponics systems, pH management is fundamental. In this kind of systems plants, fish and bacteria are cultivated. In general, the nitrogen cycle is completed inside the culture water, and the final metabolite (nitrate) is assimilated by plants. As a result, the nutrients provided by fish feed are recycled, and the negative environment impacts are diminished (Diver, 2006). But there is a problem. The optimum pH value for plants, fish and bacteria is 6, 7 and 8 respectively. There is necessary reconciling pH in this systems. Tyson (2008a,b) recommended a pH value of 8 to improve nitrogen assimilation by bacteria.

3.3 Dissolved oxygen

Mechanical aeration is by far the most common and effective way to increasing DO concentrations in ponds. In semi-intensive aquaculture, aeration is applied in case of emergency.(Boyd, 1998).

Paddlewheel aerators and propeller-aspirator-pumps are the most common aerators used in aquaculture. Aeration amounts varies from 1-2 kW ha^{-1} in extensive cultures, to 15-20 kW ha^{-1} in intensive culture of marine shrimp. For every kW of aeration extra, the gains are estimated over 500 kg fish/crustacean biomass (Boyd, 1998).

All basic types of mechanical aerators have been used in aquaculture, but vertical pumps, pump sprayers, propeller-aspirator-pumps, paddle wheels, and diffused-air systems are the most common. However, paddle wheels aerators and propeller-aspirator-pumps are the most efficient devices (Boyd, 1998).

There are some useful indications about the use of aerators. In semi-intensive culture, is common to turn on mechanical aeration at night, when is a lack of photosynthesis and the respiration rate is in a maximum value. However, is important to note that if the aerators are operated on sunny afternoons when water is supersaturated, oxygen will be lost (degassed) from the water (Tucker, 2005)

When commercial aerators are not available, a degassing column can be constructed with local available materials. Wheaton (1982) mentioned how to build a simple aeration tower. Tucker (2005) gives photos and a short description about it uses on farm systems. Theoretical background can be found in Boyd (1998), Summerfelt et al. (2000) and Vinatea & Carvalho (2007)

3.4 Nitrogen compounds

Ebeling et al. (2006) pointed out the following: for every gram of ammonia-nitrogen converted to nitrate-nitrogen, 4.18 of dissolved oxygen and 7.05 g of alkalinity (1.69 g of inorganic carbon) are consumed and 0.20 g of microbial biomass (0.105 g organic carbon) and 5.85 g of CO_2 (1.59 g inorganic carbon) are produced. So the nitrogen cycle in pond water affects the physical, chemical and biological components present in the system. To control the accumulation of every kind of components in the pond the most usual method is the addition of make-up water. However, there are alternatives to manage the levels of nitrogen compounds. Some of them are described below.

3.4.1 Ammonia

The most extended method to control ammonia inside aquacultural ponds is to keep a good feeding schedule based on nutrimental tables. If the feeding regimen in the fish farm is intensive, then pH control is also recomended. If pH can be maintained lower than 8, the toxic fraction of TAN will remain in minimal percentage (Hargreaves and Tucker, 2004).

A common method to control ammonia in recirculating aquaculture systems is a biofilter addition in the system. Biofilters design and theoretical foundations are already in literature (Bazil (2006), Drennan II et al. (2006), Eding et al. (2006), Gutierrez-Wing and Malone (2006), Kuo-Feng & Kuo-Ling (2004)). Examples of applications and affordable designs can be found in Soto-Zarazúa et al. (2010), Timmons et al. (2006), Al-Hafedh et al. (2003) and Ridha & Cruz (2001)

Another method to control ammonia is the increment of microbiological activity inside the pond (Hargreaves and Tucker, 2004). The addition of organic carbon in relation with nitrogen concentration is very useful (Crab et al., 2009). Theoretical background of this technique (also named bio-floc technology) is already available in Avnimelech (2006, 2003 and 1999).

3.4.2 Nitrite

A common practice to reduce nitrite toxicity is the elevation of chloride concentration in the culture water (Losordo et al., 1998). For this purpose, common salt (sodium chloride) is used. Calcium chloride can also be used. A high chloride:nitrate ratio of 10:1 is used to prevent brown blood disease. In catfish, for example, is recommended to maintain at least 100 ppm chloride in pond waters (Durborow et al., 1997).

When brown gill disease is not presented, a good practice is to flush water. In general, the absorption of NO_2 by bacteria is presented in tanks and ponds. To enhance its activity, biofiltration is recommended.

3.4.3 Nitrate

In general, nitrate is a non-toxic form of nitrogen in pond. However, nitrate management is usually carried out by water exchanges. Phytoplankton and bacterial uptake is another method to assimilate nitrate into cellular tissue (Gross et al., 2000). Aquaponics is another way to assimilate NO_3 into fresh, marketable plant biomass, as lettuce, tomato, basil and other crops (Graber & Junge, 2009; Rico-García et al., 2009; Savidov et al., 2005).

Denitrification is also an alternative to eliminate aquacultural nitrate. Theory and practice examples can be found in van Rijn et al., 2006.

3.5 Solids

The management of solids include feed design and management, flow regulation and separation treatment technology (Cripps and Bergheim, 2000). In small scale aquaculture, a good start point is to asses an adequate feed schedule. In our experience, we have noted that overfeeding is usual in new managers, so the detriment of the water quality is faster.

Fish also waste a lot of feed. In the case of tilapia, even 50% of their feed can be wasted (Avnimelech, 2003). So the experience of the manager is fundamental in solids control.

A complementary approach to treat solid wastes is a quickest remotion of them (Timmons et al., 1998). In general, the settleable solids can be easily remove from the culture water. If circular tanks are used, the centrifugal forces and conical bottoms must help to accumulate the solids in the center drain. (Timmons et al., 1998; Summerfelt et al., 1998). If rectangular tanks are used, is recommended to adjust length/width ratios to increased bottom velocities and reduced biosolid accumulation (Oca and Masaló, 2007). In some cases additional components can be used, like settling basins or hydroclones (Wheaton, 1982). For aquaponics, the size reduction of solid wastes from pellets to fine particles or even until basic organic compounds is desirable (Rakocy & Hargreaves, 1993).

If extensive ponds are used, the use of chemicals to enhance flocculation of organic particulate material could be useful. The addition of gypsum helps to diminish the negative electrical charges between particulate material, enhancing the flocculation and sedimentation of suspended solids. The recommended amounts are between 100 to 300 mg per liter (Hargreaves, 1999).

3.6 Carbon dioxide (CO_2)

In general terms, rarely CO_2 concentration cause problems in fish ponds with sufficient alkalinity. However, if the pond is deeper than 5 feet and poor mixing is presented, there is a risk of water stratification (Hargreaves and Brunson, 1996). Vigorous aeration can prevent stratification, and it would be helpful if CO_2 is saturated on pond water (Wurts and Durborow, 1992).

Other alternative is chemical treatment. The addition of quicklime, hydrated lime or sodium carbonate will increase the alkalinity of the pond, dropping the presence of dissolved CO_2 in water. Treatment calculations can be found on Hargreaves and Brunson, (1996). In the case of intensive systems, CO_2 could be a problem. In this case, the addition of a degassing tower is recommended (Summerfelt and Sharrer, 2004).

4. Conclusions

The processes involved in aquaculture water quality are affected by many variables and they are generally complex. However, the driving forces involved in fish farms can be controlled in order to increase fish productivity. This control involves learned skills to make the right decisions to correct present problems and to prevent new ones.

In this chapter we had discussed extensively the theoretical foundations of fish water quality, with emphasis on small-scale aquaculture correction techniques. As it can be seen, physical, chemical and biological processes involved in the pond dynamics are strongly related, and they can't be considered as isolated phenomena. The understanding of these processes can be useful to improve the final yields on the farm with the minimal waste of time, energy and money. As a result, the sustainability of the protein production can be rise and the negative environmental impact can be diminished.

5. References

Akinwole, A. O. & Faturoti, E. O. (2007). Biological performance of African catfish (*Clarias gariepinus*) cultured in recirculating system in Ibadan. Aquacultural engineering 36:18-23.

Al-Hafedh Y. S., Alam A. & Alam M. A. (2003). Performance of plastic biofilter media with different configuration in a water recirculation system for the culture of Nile tilapia (*Oreochromis niloticus*). *Aquacultural engineering* 29:139-154.

Alagaraja, K. (1991). Aquaculture productivity. On *Proceedings of the Symposium on Aquaculture Productivity*. December 1998. Hindustan Lever Research Foundation. Oxford & IBH Publishing Co. PVT. LTD. New Delhi. Sinha, V.R.P and H. C. Srivastava.

Alatorre-Jácome, O. (2010). Implementación de un sistema thermosolar para alevinaje (Nursery) de tilapia (*Oreochromis niloticus* L.). MSc. thesis dissertation. Universidad Autónoma de Querétaro, México.

Alatorre-Jácome, O., Rico-García, E., Soto-Zarazúa, G. M. García-Trejo, F. & Herrera-Ruiz, G. (2011). A thermosolar nursery for tilapia (*Oreochromis niloticus* L.). Scientific Research and Essays. In Press.

Avnimelech, Y. (2006). Bio-filters: The need for a new comprehensive approach. *Aquacultural engineering* 34:172-178.

Avnimelech,Y. (2003). Control of microbial activity in aquaculture systems: active suspension ponds. *World Aquaculture* 34(4):19-21.

Avnimelech, Y. (1999). Carbon/nitrogen ratio as a control element in aquaculture systems. *Aquaculture* 176: 227-235.

Brazil, B. L. (2006). Performance and operation of a rotating biological contactor in a tilapia recirculating aquaculture system. *Aquacultural engineering* 34:261-274.

Boyd, C. (1998) Pond water aeration systems. *Aquacultural Engineering* 18:9-40.

Brander, K. M. (2007). Global fish production and climate change. *Proceedings of the National Academy of Sciences* 104 (50): 19709–19714

Brenan, L. & Owende, P. (2010). Biofuels from microalgae — A review of technologies for production, processing, and extractions of biofuels and co-products. *Renewable and Sustainable Energy Reviews* 14:557–577.

Camargo, J. A., Alonso, A. & Salamanca, A. Nitrate toxicity to aquatic animals: a review for freshwater invertebrates. *Chemosphere* 58:1255-1267

Chen, S. & Malone, R. F. (1991). Suspended solids control in recirculating aquacultural systems. Engineering aspects of intensive aquaculture, pp. 170-186 in Proceedings from the aquaculture symposium, Cornell University, Ithaca, New York, 1991.

Chopin, T. (2003). Integrated aquaculture. *Canadian Geographic* 123(5):24

Crab, R., Kochva, M., Verstraete, W. & Avnimelech, Y. (2009). Bio-flocs technology application in over-wintering of tilapia. *Aquacultural Engineering* 40:105-112.

Cripps, S. J. & Bergheim, A. (2000). Solids management and removal for intensive land-based aquaculture production systems. *Aquacultural Engineering* 22:33-56.

Diver, S. (2006). Aquaponics - Integration of hydroponics with aquaculture. Publication No. IP163. ATTRA. National Sustainable Agriculture Information Service.

Duffie, J. A. & Beckman, W. A. (1992). Solar Engineering of thermal processes. 2th edition. John Wyley & Sons, Inc. USA.

Drennan II, D. G., Hosler, K. C., Francis, M., Weaver, D., Aneshansley, E., Beckman, G., Johnson, C. H. & Cristina, C.M. (2006). Standardized evaluation and rating of biofilters II. Manufacturer´s and user´s perspective. *Aquacultural engineering* 34:403-416

Durborow, R. M., Crosby, D. M. & Brunson, M. W. (1997a). Ammonia in fish ponds. SRAC Publication No. 463.

Durborow, R. M., Crosby, D. F. & Brunson, M. W. (1997b). Nitrite in fish ponds. SRAC Publication No. 462.

Eding, E. H., Kamstra, A., Verreth J. A.J., Huisman E. A. & Klapwijk, A. (2006). Design and operation of nitrifying trickling filters in recirculating aquaculture: A review. *Aquacultural engineering* 34:234-260.

Ebeling, J. E., Timmons, M. B. & Bisogni, J. J. (2006) Engineering analysis of the stoichiometry of photoautotrophic, autotrophic, and heterotrophic removal of ammonia-nitrogen in aquaculture systems. *Aquaculture* 257:346-358.

Eshchar, M., O., Lahav, N. Mozes, A. Peduel & B. Ron (2006). Intensive fish culture at high ammonium and low pH. *Aquaculture* 255:301-313.

Food and Agriculture Organizations of the United Nations (FAO). (2010). The State of World Fisheries and Aquaculture (SOFIA), 2010. FAO Fisheries Department, Roma.

Franck, S., von Bloh, W., Müller, C., Bondeau, A. & Sakschewski, B. (2011). Harvesting the sun: New estimations of the maximum population of planet earth. *Ecological Modelling* 222(12):2019-2026.

Fuller, R. J. (2007). Solar heating systems for recirculation aquaculture. *Aquacultural Engineering* 36:250-260.

Graber, A. & Junge, R. (2009). Aquaponic systems: Nutrient recycling from fish wastewater by vegetable production. *Desalination* 246:147-156.

Gross, A., Boyd, C. E, Wood & C. W. (2000). Nitrogen transformations and balance in channel catfish ponds. *Aquacultural Engineering* 24:1-14.

Gutierrez-Wing, M. T. & Malone, R. F. (2006). Biological filters in aquaculture: Trends and research directions for freshwater and marine applications. *Aquacultural Engineering* 34:163-171.

Hannesson, R. (2003). Aquaculture and fisheries. *Marine policy* 27:196-178.

Hargreaves, J. A. (2006). Photosynthetic suspended-growth systems in aquaculture. *Aquacultural engineering* 34:344-363

Hargreaves, J. A. & Tucker, C. S. (2004). Managing ammonia in fish ponds. SRAC Publication No. 4603. USA.

Hargreaves, J. A. (1999). Control of clay turbidity in ponds. SRAC Publication No. 460.

Hargreaves, J. A. (1998). Nitrogen biogeochemistry of aquaculture ponds. *Aquaculture* 166:181-212.

Hargreaves, J. & Brunson, M.(1996). Carbon dioxide in fish ponds. SRAC Publication No. 468.

Hovanec, T. A. & DeLong, E. F. (1996). Comparative analysis of nitrifying bacteria associated with freshwater and marine aquaria. *Applied and environmental microbiology* 62(8):2888-2896.

Jensen, F. B. (2003). Nitrite disrupts multiple physiological functions in aquatic animals. *Comparative Biochemistry and Physiology Part A* 135:9-24.

Kawarazuka, N. (2010). The contribution of fish intake, aquaculture, and small-scale fisheries to improving nutrition: A literature review. The World Fish Center Working Paper No. 2106. The World Fish Center, Penang, Malasya. 51 p.

Koning, N. B. J., Van Ittersum, M. K., Becx, G. A., Van Boekel, M. A. J. S., Brandenburg, W. A., Van Den Broek, J. A., Goudriaan, J., Van Hofwegen, Jongeneel, G., R. A.,Schiere, J. B. & Smies, M. (2008). Long-term global availability of food: continued abundance or new scarcity? *NJAS* 55(3):229-292.

Kuo-Feng, T. & Kuo-Lin W. (2004). The ammonia removal cycle for a submerged biofilter used in a recirculating ell culture system. *Aquacultural Engineering* 31:17-30.

Losordo, T. M., Masser, M. P. & Rakocy, J. E. (1999). Recirculating aquaculture tank production systems. A review of component options. SRAC publication No. 453. USA.

Losordo, T. M., Masser, M. P. & Rakocy. (1998). Recirculating aquaculture tank production systems. An overview of critical considerations. SRAC publication No. 451. USA.

Lukefar, S. (1999). Teaching international animal agriculture. *Journal of Animal Science* 77(11):3106-3113.

Masser, M. P; Rakocy, J. & Losordo, T. M. (1999). Recirculating aquaculture tank production systems. Management of recirculating systems. Sourthern Regional Aquaculture Center, Publication No. 452. USA.

Müller-Esterl, W. (2008). Bioquímica. Fundamentos para la medicina y ciencias de la visa. Editorial Reverté, S. A. España.

Mohanty, B. P., Mohanty, S., Sahoo, J. K. & Sharma, A. P. (2010). Climate change: impacts on fisheries and aquaculture. In Simard, S. (ed.) Climate change and variability InTech Publishing ISBN 978-953-307-144-2

Naylor, R. L., Hardy, R.W., Bureau, D.P., Chiu, A., Elliot, M., Farrell, A. P., Forster, I.,Gatlin, D.M., Goldburg, R.J., Hua, K. & Nichols, P. D. (2009). Feeding aquaculture in an era of finite resources. *Proceedings of the National Academy of Sciences* 106 (36):15103-15110.

Naylor, R. L., Goldburg, R.J., Primavera, J. H., Kautsky, N., Beveridge, M. C. M., Clay, J., Folke, C., Lubchenco, J., Mooney, H., & Troell, M. (2000). Effect of aquaculture on world fish supplies. *Nature* 405(29):1017-1024.

Oca, J., & Masaló, I. (2007). Design criteria for rotating flow cells in rectangular aquaculture tanks. *Aquacultural Engineering* 36:36-44.

Olvera-Olvera, C., Olvera-González, J. E., Mendosa-Jasso, J., Peniche-Vera, R., Castañeda-Miranda, R. & Herrera-Ruiz, G. (2009). Feed dosage and ammonium control device base on C/N ratio for a zero-discharge system. International Journal of Agriculture & Biology 11(2).170-177.

Panjaitan, P. (2010). Shrimp culture of *Penaeus monodon* with zero water exchange model (ZWEM) using molasses. *Journal of Coastal Development* 14(1):35-44

Rakocy, J. E., & Hargreaves, J. A. (1993). Integration of vegetable hydroponics with fish culture: A review. *In* Wang, Jaw-Kai. Techniques of modern aquaculture. Proceedings of an Aquacultural Engineering Conference. 21 – 23 June, 1993. Spokane, Washington .pp. 112-136. Published by American Society of Agricultural Engineerings.

Rico-García, E.; Casanova-Villareal, V. E., Mercado-Luna, A., Soto-Zarazúa, G. M., Guevara-González, R. G., Herrera-Ruiz, G., Torres-Pacheco, I., Velázquez-Ocampo, R. V. (2009). Nitrate content on summer lettuce production using fish culture water. *Trends in Agriculture Economics* 2(1):1-9.

Ridha, M. T. & Cruz, E. M. (2001). Effect of biofilter media on wáter quality and biological performance of the Nile tilapia *Oreochromis niloticus* L. reared in a simple recirculating system. *Aquacultural Engineering* 24:157-166

Rumei, W., Setian, F., Guo, T., Lizhong, F. & Xiaoshuan (2003). Evaluation of the aquaculture pond water quality. ASAE Annual International Meeting. Nevada, USA. Paper ASAE 031298.

Savidov, N. A., Hutchings, E., Rakocy, J. E. (2005). Fish and plant production in a recirculating aquaponic system: A new approach to sustainable agriculture in Canada. *Acta horticulturae* 742:209-222.

Seginer, I. & Halachmi, I.(2008). Water heating to enhance fish growth. Is it justified? Proceedings AgEng2008 (CD), Paper OP-630 (1131237).

Summerfelt, S. T. & Sharrer, M. J. (2004). Design implication of carbon dioxide production within biofilters contained in recirculating salmonid culture systems. *Aquacultural engineering* 32:171-182.

Summerfelt, S. T., Vinci, B. J. & Piedrahita, R. H. (2000). Oxygenation and carbon dioxide control in water reuse systems. *Aquacultural engineering* 22:87-108.

Summerfelt, S. T., Timmons, M. B., Watten, B. J. (1998). Culture tank designs to increase profitability. *In* Libey, G. S. and Timmons, M. B. (1998) Proceedings of the Second International Conference on Recirculating Aquaculture. July 16-19, Roanoke, Virginia.

Timmons, M. B.; Holder, J. L., Ebeling, J. B. (2006). Application of microbead biological filters. *Aquacultural engineering* 34:332-343.

Timmons, M. B.; Ebeling, J. M., Wheaton, F. W., Summerfelt, S. T., Vinci, B. J. (2002). Recirculating aquaculture systems. Cayuga Aqua Ventures Inc. 2th. Edition. USA.

Timmons, M. B.; Summerfelt, S. T., Vinci, B. J. (1998). Review of circular tank technology and management. *Aquacultural Engineering* 18:51-69.

Soto-Zarazúa, M. G., Rico-García, E. & Toledano-Ayala, M. (2011). International Journal of the Physical Sciences 6(5):1039-1044.

Soto-Zarazúa, M. G., Herrera-Ruiz, G., Rico-García, E., Toledano-Ayala, M., Peniche-Vera, R., Ocampo-Velázquez, R. & Guevara-González, R. (2010). Development of efficient recirculation system for Tilapia (Oreochromis niloticus) culture using low cost materials. *African Journal of Biotechnology* 9(32):5203-5211.

Tucker, C. (2005). Pond aeration. Sourthern Regional Aquaculture Center, Publication No. 3700. USA.

Tyson, R. V., Simonne, E. H., Treadwell, D. D., Davis, M. & White, J. M. (2008a). Effect of water ph on yield and nutritional status of greenhouse cucumber grown in recirculating hydroponics. *Journal of plant nutrition* 31(11):2018-2030.

Tyson, R. V., Simonne, E. H., Treadwell, D. D., White, J.M. & Simone, A. (2008b). Reconciling pH for ammonia biofiltration and cucumber yield in a recirculating aquaponic system with perlite biofilters. *HortScience* 43(3):719-724.

Van Gorder, S. (2003). Small-scale aquaculture and aquaponics. The new and the nostalgic. *Aquaponics Journal* 7(3):14-17.

Van Rijn, J., Tal, J. & Schreier, H. J. (2006). Denitrification in recirculating systems: Theory and applications. *Aquacultural Engineering* 34:364-376.

Vinatea, L. & Carvalho, J. W. Influence of water salinity on the SOTR of paddlewheel and propeller-aspirator-pump aerators, its relation to the number of aerators per hectare and electricity costs. *Aquacultural engineering* 37 (2): 73-78.

Wheaton, F. W. (1982). Acuacultura. Diseño y construcción de sistemas. AGT Editor, S. A. México, D. F.

Wurts, W. A. & Durborow, R. M. (1992). Interactions of pH, carbon dioxide, alkalinity and hardness in fish ponds. Southern Regional Aquaculture Center, Publication No. 464. USA.

Yi, Y. & Kwei-Lin, C. & Diana, J. S. (2003). Techniques to mitigate clay turbidity problems in fertilized earthen fish ponds. *Aquacultural Engineering* 27:39-51.

7

Integrated Multitrophic Aquaculture: Filter Feeders Bivalves as Efficient Reducers of Wastes Derived from Coastal Aquaculture Assessed with Stable Isotope Analyses

Salud Deudero[1], Ariadna Tor[1], Carme Alomar[1],
José Maria Valencia[2], Piluca Sarriera[1] and Andreu Blanco[3]
[1]Centro Oceanográfico de Baleares, Instituto Español de Oceanografía, Palma
[2]Laboratorio de Investigaciones Marinas y Acuicultura, Andratx
[3]Instituto de Investigaciones Marinas de Vigo, CSIC, Vigo
Spain

1. Introduction

Aquaculture industry has rapidly increased from 20 million tons in the 1990s to 68.4 million tons in 2008, overcoming the 67.0 million tons of extractive fisheries, with an approximated global value of 84,791 million Euros (FAO, 2010). The Mediterranean Sea is an important producer of intensive open water fish culture, especially important are industries from Spain, France, Italy and Greece, which generated 2,133 million Euros in 2008 (FAO, 2008).

Parallel to aquaculture industry development, knowledge of its effects on the surrounding environment is increasing. Several studies have been conducted to assess the environmental impact (Holmer et al., 2008; Yokoyama et al., 2006; Vizzini & Mazzola, 2004) due to aquaculture. Organic enrichment is recognized as the most important problem associated to marine aquaculture (Mirto et al., 2010; Troell et al., 2003) as a direct result of the release of dissolved and particulate nutrient loads, especially organic phosphorous and nitrogen in the form of ammonia that might easily induce eutrophication (Karakassis et al., 2000). Uneaten pellet and fresh food supplied at fish cages, together with the excretion products from cultured fishes are the source of the nutrient loads released (Cheshuk et al., 2003).

Whereas the dissolved compounds are easily dispersed and diluted in the water column, the particulate compounds sink to the sea floor, causing severe modifications of the physical and chemical characteristics of the sediment and the community dynamics of marine seagrass and benthic fauna (Brown et al., 1987; Karakassis et al., 2000; Mente et al., 2006). The severity and extend of the environmental impact depends on a large number of factors, such as local hydrodynamics, water depth and total output from fish farms. Nevertheless, it has been detected organic enrichment in the sediment up to 1000 m from fish cages (Sarà et al., 2004).

The effects of the organic enrichment are widely studied in the Atlantic and the Pacific Ocean, especially in relation to the salmon industry (Brown et al., 1987; Cheshuk et al., 2003;

Weston, 1991), however, knowledge of fish farming impacts in the Mediterranean Sea is scarce (Sarà et al., 2004, 2006; Vizzini & Mazzola, 2004), which have different hydrographic characteristics, such as shallow waters and low current velocity areas, and are characterized by oligotrophic waters.

1.1 Integrated multitrophic aquaculture

The growing concern on aquaculture's environmental impact has led to an increasing research into feed formulations and digestibility, better conversion efficiency and improved management (Skalli et al., 2004; Troell et al., 2003). Since the last decade, the emphasis has been placed on the practice of integrated multi-trophic aquaculture (IMTA), with a potential to mitigate some of the environmental problems associated with mono-specific aquaculture (Soto et al., 2008). Originally used in freshwater practices, it involves the culture of two or more species from different trophic level; generally finfish being simultaneously cultured with both organic and inorganic extractive species, such as shellfish and seaweeds, respectively; in which by-products from one species are recycled to become inputs for another. Thus, the organic matter released in aquaculture systems might represent a source of available food for filter-feeding organisms, such as bivalves, reducing its impact on the environment (Shpigel et al., 1991) and can represent a potential economic income.

Filter feeders bivalves are essentially generalist consumers, and it has been demonstrated that they can exploit organic matter from several sources (autochthonous, allochthonous or anthropogenic), as a function of its availability (Stirling & Okumus, 1995). In a conceptual open water integrated aquaculture, filter feeder bivalves are cultured adjacent to fish floating cages, reducing nutrient loadings by filtering and assimilating particulate wastes (uneaten food and faeces) as well as phytoplankton. In this way, bivalves would perform as biological filters. Previous studies have determined that bivalves can be successfully incorporated into integrated multitrophic aquaculture systems, based on the increased growth displayed and the feeding efficiency on pellet feed and fecal products (Mazzola & Sarà, 2001; Reid et al., 2010).

1.2 Stable isotopes as tracers of matter fluxes

The use of stable isotopes in biogeochemistry and marine ecosystem analysis is increasing rapidly, especially in assessing trophic relationships and pathways of energy flow in food webs (Bergamino et al., 2011). Traditionally, the origin and fate of organic matter in the marine environment have been investigated through different approaches, such as lipid biomarkers or chlorophyll pigments. In filter feeder bivalves, it has been evaluated with gut content or fatty acid analyses, representing an instant snapshot of food ingested by bivalves. This limitation could be solved using stable isotope analyses, as the evaluation of food sources is based on assimilated instead of ingested food, which represents a time-integrated food utilization.

Stable isotope analysis is based on the assimilation of ^{13}C and ^{15}N from sources, with a slight enrichment of heavier isotopes (^{13}C and ^{15}N) as lighter isotopes (^{12}C and ^{14}N) are used in metabolism. The enrichment between prey and consumer tissues has been considered to be consistent across species, however, it has been demonstrated that it depends on the assimilation process and it has been set at 1‰ for ^{13}C and 3-4‰ for ^{15}N

Integrated Multitrophic Aquaculture: Filter Feeders Bivalves as Efficient Reducers of Wastes Derived from
Coastal Aquaculture Assessed with Stable Isotope Analyses

121

(De Niro & Epstein, 1978; McCutchan et al., 2003). Since the ratio of $^{12}C/^{13}C$ isotopes changes little through the food web, this ratio is commonly used to distinguish between carbon sources, typically C_3 or C_4 and pelagic or benthic. In contrast, consumers showed a large amount of enrichment in nitrogen, thus, the $^{14}N/^{15}N$ ratio is commonly used to estimate trophic positions. Integrated multitrophic aquaculture involves species with different trophic strategies which can be positioned in the food web as a function of the ^{13}C and ^{15}N stable isotope signatures.

1.3 Objectives

To decipher whether integrated multitrophic aquaculture is an effective method for minimizing and reducing waste inputs into the Mediterranean coastal ecosystem we have designed a multitrophic integrated system involving filter feeders bivalves (*galloprovincialis (Lamarck, 1819)* and *Chlamys varia (Linnaeus, 1758))*, and fin fish *Argyrosomus regius (Asso, 1801)*. Through determination of carbon and nitrogen stable isotopes analyses to the several organisms and trophic strategies, the following aims will be accomplished:

i.   Define the isotopic composition of the trophic food web of integrated multitrophic aquaculture in relation to two reference stations
ii. Study the temporal variability in waste matter fluxes at the different treatments, hence annual and seasonal variability
iii. Calculate the relative contributions of wastes (fresh food, pellets, plankton) to the fish farmed and to the filter feeders at the IMTA

2. Methods

2.1 Experimental design for IMTA

The study was conducted during years 2008, 2009 and 2010 at a research experimental station (LIMIA), on the SW coast of Mallorca, in Andratx Bay (39° 32' 38.13" N, 2° 22' 51.08" E) (Figure 1). The installations included 6 floating cages for fish reproductive stages in the middle of the bay, with cultured *Argyrosomus regius*, with a total fish stock of 12-15 t year $^{-1}$. Water depth ranged between 5 to 8 meters. Sediments surrounding fish cages were mostly unvegetated, except for disperse patches of *Caulerpa prolifera (Forsskal) J.V.Lamouroux, 1809* within tens of meters distance. Current velocity below fish cages varied between 0 to 10 cm/s, with higher prevalence of very low current speeds (0 to 5 cm/s) (own data). Feed regime consisted in a combination of commercial dry pellets, CV4, Mar-9 vitalis repro and Gemma (Trouw S.A), hereafter called pellet 1, pellet 2 and pellet 3, respectively; and fresh food based on fishes *Sardina pilchardus (Walbaum, 1792)* (34.48 %), *Spicara smaris (Linnaeus, 1758)* (24.14 %), *Trachurus trachurus (Linnaeus, 1758)* (31.03 %) and squid *Loligo vulgaris Lamarck, 1798* (10.34 %) on average ten times per month, with 10kg supplied each time.

Two reference sites were selected to evaluate the effects of aquaculture wastes in the adjacent environment. An external site (control 1) (39° 32' 29.37" N, 2° 22' 56.45" E) located approximately 350 m away from the fish cages within the Andratx Bay, was selected to study the influence of the natural variability in the bay. The second site, (control 2) was located 21 nautical miles away from fish cages (39° 28' 4" N, 2° 42' 53" E) in order to compare the isotopic values of the different components of the food web without the fish farm influence.

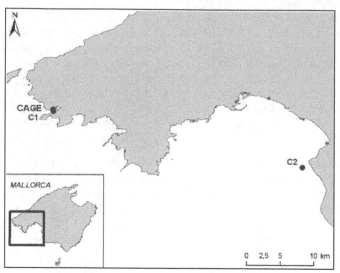

Fig. 1. Map of the SW coast of Mallorca. Solid circles indicate sampling sites (Cage: cage site, C1: control 1, C2: control 2).

Filter feeder bivalves *Mytilus galloprovincialis* and *Chlamys varia* were chosen for this study as they are native to the region, fast growing and commercially viable. They were collected from the harbor area in Andratx Bay between April and July 2007 and transferred to polyethylene plastic bags (0.70 x 0.40 m and 0.009 m mesh diameter) with a density of 250 *Mytilus galloprovincialis*/bag (size between 44 and 72 mm) and 100 *Chlamys varia*/bag (size between 19 to 42 mm). Forty plastic bags were placed hanging on fish cages, and the same amount was simultaneously attached to the rope buoys in control 1 and control 2.

2.2 Sampling and data collection

2.2.1 Cage site

In the experimental site (hereafter called cage site), approximately 10 adult specimens of *Argyrosomus regius*, and 6 samples of *Mytilus galloprovincialis* and *Chlamys varia* (Linnaeus, 1758) were sampled seasonally. Additionally, potential food sources derived from aquaculture wastes as fish faeces and pellet and fresh food were sampled; as well as the other components of the marine food web, phytoplankton, zooplankton, particulate organic matter (POM) and sedimentary organic matter (SOM). Moreover, 5 sedimentivorous echinoderms *Holothuria forskalii* Delle Chiaje, 1823 were also collected.

Fish faeces were collected using pipettes after transferring adult *Argyrosomus regius* in a conical tank for an overnight evacuation. Individuals were then sacrificed in order to extract the muscle tissue from the left side of the posterior region to the dorsal fin using stainless-steel tools, then, samples were cleaned thoroughly and rinsed with distilled water. Phytoplankton and zooplankton samples were collected using a plankton net (100 µm mesh size, 30 cm diameter) and a WP2 plankton net for zooplankton (250 µm mesh size, 45 cm diameter), respectively, which were towed through the water subsurface for 15 minutes at a velocity of 2 knots.

Integrated Multitrophic Aquaculture: Filter Feeders Bivalves as Efficient Reducers of Wastes Derived from
Coastal Aquaculture Assessed with Stable Isotope Analyses

123

Particulate organic matter, was sampled by vacuum filtering 10 L of seawater, collected using a 5 L Niskin bottle (General Oceanic, USA), through a pre-combusted (450°C, 4 hours) Whatman GF/F glass microfiber filter, 1 m above the floating cages while in control sites, it was collected at the same depth as the filter feeders bivalves polyethylene bags. Sedimentary organic matter was collected by SCUBA divers from a quadrant surface (400 cm^2) using hand cores.

2.2.2 Control sites

Similarly, in control sites, 6 specimens of filter feeder bivalves *Mytilus galloprovincialis* and *Chlamys varia* were sampled. In control 1, *Chlamys varia* could not be sampled since the polyethylene bags were lost after severe storm periods. Moreover, phytoplankton, zooplankton, particulate organic matter (POM) and sedimentary organic matter (SOM) were sampled seasonally and 5 samples of sedimentivorous echinoderms *Holothuria forskalii* were sampled annually following the same procedure as in cage site (Table 1). All samples were frozen immediately after sampling and kept at -20°C till further processing.

2.3 Stable isotope analyses

Samples were dried in an oven at 60°C for 24h and subsequently grounded to a fine powder using a mortar and pestle. A minimum of two replicates for each sample was analyzed for $\delta^{13}C$ and $\delta^{15}N$ isotopic signatures, except for commercial dry pellet feed which had a stable and controlled composition (Trouw S.A). SOM samples for $\delta^{13}C$ isotopic analysis were acidified by adding 2N HCl (Carabel et al., 2006), while for $\delta^{15}N$ analysis, non- acidified replicates were used. From each sample 2 ± 0.1 mg of dry weight was placed in tin cups to determine the stable isotope ratios of carbon and nitrogen. The analyses were run at the SCTI (Scientific-Technical Services) from the Balearic Islands' University using a continuous flow mass spectrometer (Thermo Finnegan Delta x-plus). Reference standards were Vienne Pee Dee Belemnite (VPDB) for C and atmospheric nitrogen for N. One sample of an internal reference, Bovine Liver Standard (1577b) (U.S. Department of Commerce, National Institute of Standards and Technology, Gaithersburg, MD 20899), was analyzed every eight samples in order to calibrate the system and compensate the drift over time. The analytical precision of the stable isotope analyses was based on the standard deviation of the BSL samples; these deviations were 0.08 ‰ for $\delta^{13}C$ and 0.09 ‰ for $\delta^{15}N$. Isotope ratios were expressed in $\delta^{13}C$ and $\delta^{15}N$, with units of ‰, according to the following equation:

$$\delta^{13}C \text{ or } \delta^{15}N = [(R_{sample} / R_{reference}) - 1] \times 1000$$

Where R is the corresponding $^{13}C/^{12}C$ or $^{15}N/^{14}N$ ratio.

2.3.1 Statistical analyses

A distance-based permutational analysis of variance (PERMANOVA) on 999 permutations was employed to test the hypothesis that there were no differences in the isotopic signature ($\delta^{13}C$ and $\delta^{15}N$) of filter feeders bivalves between treatments and sampling periods. The factors Treatment (with three levels: cage, control 1 and control 2), Year (with three levels: 2008, 2009 and 2010) and Season nested in Year (with three factors: spring, summer and autumn) were treated as fixed. Additionally, a distance-based test for homogeneity of

Sample	Treatment	2008	2009	2010	Perodicity
Fin Fish					
Argyrosomus regius	cage	x	x	x	Seasonal
Filter feeders					
Mytilus galloprovincialis	cage	x	x	x	Seasonal
Mytilus galloprovincialis	control 1	x	x	x	Seasonal
Mytilus galloprovincialis	control 2	x	x	x	Seasonal
Chlamys varia	cage	x	x	x	Seasonal
Chlamys varia	control 1	–	–	–	Seasonal
Chlamys varia	control 2	–	x	x	Seasonal
Sedimentivorous					
Holothuria forskalii	cage	–	x	x	Anual
Holothuria forskalii	control 1	–	x	x	Anual
Holothuria forskalii	control 2	–	x	x	Anual
Food sources					
Pellet food					
Pellet 1 (CV4)	cage	x	–	–	Anual
Pellet 2 (Mar-9 vitalis repro)	cage	x	–	–	Anual
Pellet 3 (Gemma)	cage	x	–	–	Anual
Fresh food					
Loligo vulgaris	cage	x	x	–	Seasonal
Sardina pilchardus	cage	x	x	–	Seasonal
Spicara smaris	cage	x	x	–	Seasonal
Trachurus trachurus	cage	x	x	–	Seasonal
Organic matter sources					
Phytoplankton	cage	–	x	x	Seasonal
Phytoplankton	control 1	–	x	x	Seasonal
Phytoplankton	control 2	–	x	x	Seasonal
Zooplankton	cage	x	x	x	Seasonal
Zooplankton	control 1	x	x	x	Seasonal
Zooplankton	control 2	x	x	x	Seasonal
POM	cage	x	x	x	Seasonal
POM	control 1	x	x	x	Seasonal
POM	control 2	x	x	x	Seasonal
SOM	cage	x	x	x	Seasonal
SOM	control 1	x	x	x	Seasonal
SOM	control 2	x	x	x	Seasonal

Table 1. List of samples collected during the three years of the study in each location with its periodicity. (X: sample collected and -: sample not collected).

Integrated Multitrophic Aquaculture: Filter Feeders Bivalves as Efficient Reducers of Wastes Derived from
Coastal Aquaculture Assessed with Stable Isotope Analyses

125

dispersions (PERMDISP) was used to analyze the multivariate dispersion for each treatment at all sampling sites. All statistical computations were performed using the statistical package PRIMER® 6.0 software. The permutational analysis (PERMANOVA) was not performed on $\delta^{13}C$ and $\delta^{15}N$ sediment data in years 2009 and 2010 because, as a consequence of the low carbon and nitrogen content in the samples, some replicates did not reach the mass requirements and could not be analyzed in the mass spectrometer.

2.3.2 Partial contributions of aquaculture wastes at the studied food webs

Stable isotope mixing models are an increasingly common approach in environmental sciences. They are used to determine the proportional contribution of sources to a mixture based on their respective isotope signatures (Phillips & Gregg, 2001). Amongst its application we can find the estimation of the relative importance of food sources to animal diets, pollution sources to air or water bodies and carbon sources to soil organic matter (Michener & Lajtha, 2007).

In this study, we applied the SISUS Bayesian Mixing model (Stable Isotope Sourcing using Sampling) to quantify the feasible contributions of the potential organic matter sources (phytoplankton, zooplankton, POM, faecal material, pellet and fresh food) to the filter feeders bivalves' diet, based on the analyzed stable isotope ratios. Considering the specific isotope enrichment between prey and consumer tissues, a discrimination of 0.3 ‰ for carbon was assumed for POM and filter feeders bivalves; while for muscle tissue samples of *Argyrosomus regius* a 1.3 ‰ was applied. Previous studies have shown that trophic fractionation is much larger for $\delta^{15}N$ than for $\delta^{13}C$, so a correction of 2.3 ‰ per trophic level was applied (McCutchan et al., 2003).

The model is based on a modification of the Isosource programme (Phillips & Gregg, 2001), which creates every possible combination of source proportions and compares these predicted mixtures signatures with the observed mixtures signatures. If they are equal or within some small tolerance range, this combination represents a feasible solution. The SISUS software is available for public use at http://statacumen.com/sisus/.

3. Results

3.1 Isotopic composition of the marine food web components at IMTA

The carbon and nitrogen isotopic composition of the marine food components analyzed in the three years of study in cage and control sites are summarized in Figures 2, 3 and 4.

3.1.1 Potential organic matter sources and environmental elements

Commercial pellet food supplied in the fish farm had similar isotopic values of $\delta^{13}C$ (between -20.78 ‰ to -21.87 ‰), with an average value of -21.17 ± 0.09 ‰; however, they differed greatly in $\delta^{15}N$ values (between 14.93 ‰ and 6.20 ‰), with pellet 2 being the most enriched source in ^{15}N, in contrast to pellet 1 which was the most depleted nitrogen source. Fresh food were ^{13}C- enriched compared to pellet food, on average 2 ‰ (mean -19.1 ± 0.19 ‰), in contrast, $\delta^{15}N$ values were relatively similar. Considering all individual sources, fresh food sources had similar values of $\delta^{13}C$ (between -18.82 ‰ to -19.19 ‰), but they showed differences in $\delta^{15}N$ values, ranging between 7.92 ‰ to 12.46 ‰, with *Loligo vulgaris* as the most enriched source in ^{15}N and *Spicara smaris* the most depleted.

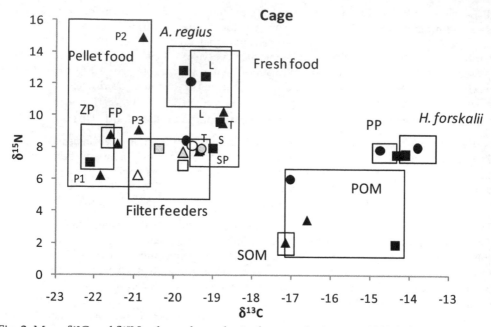

Fig. 2. Mean δ¹³C and δ¹⁵N values of samples in the cage site in years 2008 (▲), 2009 (■) and 2010 (●). ZP: zooplankton; PP: phytoplankton; FP: *Argyrosomus regius* fish faeces; S: *Sardina pilchardus*; SP: *Spicara smaris*, T: *Trachurus trachurus*; L: *Loligo vulgaris*; P1: pellet 1; P2: pellet 2; P3: pellet 3; POM: particulate organic matter and SOM: sedimentary organic matter. Filter feeders include *Chlamys varia* (grey symbol) and *Mytilus galloprovincialis* (white symbol). *Sardina pilchardus*, *Spicara smaris*, *Trachurus trachurus* and *Loligo vulgaris* are classified as fresh food.

Throughout the sampling period, *Argyrosomus regius'* carbon isotopic signature was, on average, -19.06 ± 0.64 ‰, (ranging between -18.72 ‰ to -19.77 ‰), while for nitrogen it was, on average, 11.45 ± 1.12 ‰, (ranging between 10.24 ‰ to 12.83 ‰). Concerning the isotopic-step enrichment, the value of fish faeces was ¹³C - depleted compared to the muscle isotopic signature, with a mean decrease of 1.61 ± 0.55 ‰. For ¹⁵N, the depletion was more pronounced, with a mean decrease of 3.34 ± 1.34 ‰.

Regarding the phytoplankton fraction, in the three years of study, phytoplankton from cage site exhibited similar values of δ¹³C and δ¹⁵N, with mean values of -14.54 ± 0.29 ‰ and 7.70 ± 0.22 ‰, respectively. Values in control sites showed little variability, with mean values of -15.04 ± 0.41 ‰ and 7.10 ± 1.06 ‰ for δ¹³C and δ¹⁵N in control 1; and -19.44 ± 0.44 ‰ and 5.25 ± 1.25 ‰ for δ¹³C and δ¹⁵N in control 2, respectively. For zooplankton, mean values in cage site were -21.13 ± 1.27 ‰ for δ¹³C and 8.06 ± 0.91 ‰ for δ¹⁵N. Values from control 1 showed more variability; with mean values of -18.49 ± 3.56 ‰ for δ¹³C and 7.97 ± 0.91‰ for δ¹⁵N, while in control 2 values were of -17.39 ± 0.37 ‰ and 5.91 ± 1.30 ‰ for δ¹³C and δ¹⁵N, respectively. Comparing δ¹³C and δ¹⁵N values from cage site relative to control sites, phytoplankton from cage site showed a slightly enrichment in ¹³C relative to control 1, on average 0.59 ‰; differences were more pronounced relative to control 2, with an average

Integrated Multitrophic Aquaculture: Filter Feeders Bivalves as Efficient Reducers of Wastes Derived from Coastal Aquaculture Assessed with Stable Isotope Analyses

127

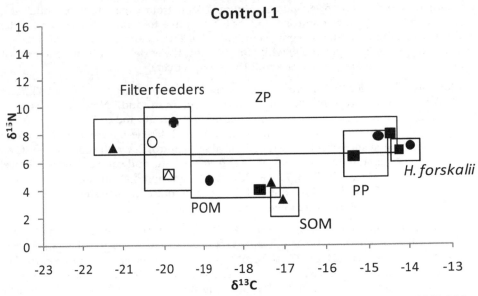

Fig. 3. Mean δ13C and δ15N values of the different samples in control 1 in years 2008 (▲), 2009 (■) and 2010 (●). ZP: zooplankton; PP: phytoplankton; POM: particulate organic matter and SOM: sedimentary organic matter. Filter feeders include *Mytilus galloprovincialis* (white symbol).

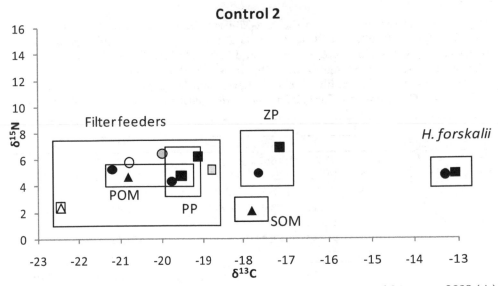

Fig. 4. Mean δ13C and δ15N values of the different samples in the control 2 in years 2008 (▲), 2009 (■) and 2010 (●). ZP: zooplankton; PP: phytoplankton; POM: particulate organic matter and SOM: sedimentary organic matter. Filter feeders include *Mytilus galloprovincialis* (white symbol) and *Chlamys varia* (grey symbol).

enrichment of 4.38 ‰. For δ15N values, phytoplankton from cage site, was enriched on average a 1.35 ‰ relative to control 1 and 1.56 ‰ relative to control 2. The same pattern existed for zooplankton, with an average enrichment of 2.63 ‰ and 3.73 ‰ in 13C relative to control 1 and control 2, respectively; while for nitrogen, the enrichment relative to control 1 and control 2 was on average, 0.09 ‰ and 2.15 ‰, respectively.

Particulate organic matter (POM) from cage site showed mean values of -15.98 ± 1.43 ‰ and 3.78 ± 2.06 ‰ for δ13C and δ15N, showing an enrichment in 13C and 15N relative to control 1 and control 2, with average values of 1.93 ‰ and 0.62 ‰ in 13C and 0.62 ‰ and 1.10 ‰ in 15N, respectively. As in phytoplankton and zooplankton, differences were more pronounced when comparing cage site with the distant control site, control 2. On the contrary, δ13C and δ15N sedimentary organic matter (SOM) values from cage site and control 1 were relatively similar, with mean values of -17.13 ‰ and 2.04 ‰ for δ13C and δ15N in cage site and -17.04 ‰ and 3.34 ‰ for δ13C and δ15N in control 1. Relative to control 2, cage site was enriched in 13C a 0.67 ‰ while for 15N the enrichment was only 0.06 ‰

Sedimentivorous echinoderm *Holothuria forskkalii* showed mean values of -13.95 ± 0.28 ‰ and 7.77 ± 0.56 ‰ for δ13C and δ15N, respectively in cage site. Average values in control 1 were -14.10 ± 0.64 ‰ and 7.04 ± 0.27 ‰ for δ13C and δ15N, respectively and in control 2 of -13.19 ± 0.62 ‰ and of 4.82 ± 0.65 ‰ for δ13C and δ15N, respectively. Comparing cage site with control sites, an enrichment of 0.79 ‰ in 13C relative to control 2 was observed, while control 1 values were depleted in a 0.15 ‰ for 13C. For 15N, it was observed an enrichment of 0.73 ‰ and 2.93 ‰ relative to control 1 and 2, respectively.

3.1.2 Filter feeder bivalves and temporal variability in isotopic signals among treatments

Filter feeders bivalves collected in the cage site exhibited similar isotopic values, ranging from -20.73 ‰ to -20.90 ‰ for 13C and 6.26 ‰ to 6.36 ‰ for 15N in *Mytilus galloprovincialis*, and between -19.41 ‰ to -19.76 ‰ for 13C and 7.33 ‰ to 7.64 ‰ for 15N in *Chlamys varia*. Statistical analyses showed significant differences between cage and control treatments in the three years of study with filter feeder bivalves from fish cages enriched in both 13C and 15N (PERMANOVA, p<0.05) (Table 2 and 3). *Mytilus galloprovincialis* of cage site showed an

| | $\delta^{13}C$ | | | | $\delta^{15}N$ | | | |
Source of variation	df	SS	MS	Pseudo-F	df	SS	MS	Pseudo-F
Tr	2	18 .07	9 .0352	20 .743*	2	17 .327	8 .6634	19 .53*
Ye	2	10 .524	5 .2618	12 .08*	2	2 .1712	1 .0856	2 .44
Se(Ye)	4	17 .003	4 .2509	9 .75*	4	31 .434	7 .8584	17 .72*
TrxYe	1	1 .6363	1 .6363	3 .75	1	8 .6544	8 .6544	19 .51*
TrxSe(Ye)	1	3 .9901	3 .9901	9 .16*	1	0 .47796	0 .47796	1 .078
Res	107	46 .607	0 .43558		107	47 .442	0 .44338	
Total	117	93 .894			117	135 .58		

Table 2. Results of the 3-way PERMANOVA, mixed design, with "Treatment" (Tr), "Year" (Ye) and "Season"(Se) as fixed factors. The analysis is based on the modified Euclidean distance dissimilarity of the *Mytilus galloprovincialis* isotopic signatures of δ13C and δ15N (999 permutations). The studied factors were 3 treatments (cage, control 1 and control 2), three years (2008, 2009 and 2010) and three seasons (spring, summer and autumn). *p<0.05

Integrated Multitrophic Aquaculture: Filter Feeders Bivalves as Efficient Reducers of Wastes Derived from
Coastal Aquaculture Assessed with Stable Isotope Analyses

129

	$\delta^{13}C$				$\delta^{15}N$			
Sources of variation	df	SS	MS	Pseudo-F	df	SS	MS	Pseudo-F
Tr	2	17.403	8.7017	16.71*	2	22.21	11.105	37.24*
Ye	2	29.116	14.558	27.97*	2	2.8058	1.4029	4.70*
Se(Ye)	4	24.883	6.2208	11.95*	4	40.425	10.106	33.89*
TrxYe	0	0		No test	0	0		No test
TrxSe(Ye)	1	0.099525	0.099525	0.19	1	1.0493	1.0493	3.51
Res	83	43.2	0.52048		83	24.745	0.29814	
Total	92	88.919			92	92.935		

Table 3. Results of the 3-way PERMANOVA, mixed design, with "Treatment" (Tr), "Year" (Ye) and "Season"(Se) "as fixed factors. The analysis is based on the modified Euclidean distance dissimilarity of the *Chlamys varia* isotopic signatures of $\delta^{13}C$ and $\delta^{15}N$ (999 permutations). The studied factors were 3 treatments (cage, control 1 and control 2), three years (2008, 2009 and 2010) and three seasons (spring, summer and autumn). *p<0.05

enrichment of 0.52 ‰ in ^{13}C and 0.75 ‰ in ^{15}N compared to bivalves from control 1; greatest differences were found when comparing cage site with control 2, with an enrichment of 0.75 ‰ in ^{13}C and 1.69 ‰ in ^{15}N (p<0.05) (Figure 5 and 6). *Chlamys varia* from cage site showed a large amount of enrichment compared to control 2, 1.11 ‰ and 1.71 ‰ in ^{13}C and ^{15}N, respectively (Figure 7 and 8).

3.2 Contribution of organic and inorganic matter sources to different consumers of the multitrophic system

According to the bayesian mixing model, the main food source for the fish *Argyrosomus regius* was pellet food, with a mean global contribution of 80.43 % (from pellet 1, 2 and 3), while the remaining 19.57 % corresponded to fresh food. Contributions varied annually, 67.7 %, 91.53 % and 82 % for pellet food and 32.3 %, 8.5 % and 18 % for fresh food in years 2008, 2009 and 2010, respectively (Figure 9). In the first year of study, pellet 1 was the main pellet food source, while *Sardina pilchardus* and *Spicara smaris* were the main fresh food sources; the main pellet food source changed to be pellet 2 the following year, while in 2010 the main source was again pellet 1; *Sardina pilchardus* and *Spicara smaris* remained the main fresh food sources along the remaining years of study.

In filter feeder bivalve *Chlamys varia*, pellet food was the main food source, with a mean contribution of up to 62.68 % of their isotopic composition in the three years of study, followed by fresh food, POM, fish faeces and phytoplankton, with a mean contribution of 18.28 %, 11.26 %, 5.18 % and 2.58 %, respectively. Contributions of the five organic matter sources varied both seasonally and annually (Figure 10), however, aquaculture derived products, as pellet and fresh food and fish faeces remained the main food sources along the three years of study varying from 70.03 % in spring 2009 to the maximum of 93.81 % in autumn 2010. Regarding the seasonal variation, there was a fluctuation in the contribution of pellet food, with a medium contribution in spring, fewer in summer and a higher contribution in autumn. Summer 2009 showed a different pattern with a contribution of 81.74 % of pellet food. Regarding pellet food sources, pellet 1 was the main source in the three years of study, with a contribution fluctuating between 18.72 % and 81.74 %, while for fresh food the main contribution was from *Sardina pilchardus* and *Spicara smaris*. Particulate organic matter (POM) also showed variation, with the lowest contribution in autumn 2009 (4.24 %) and the

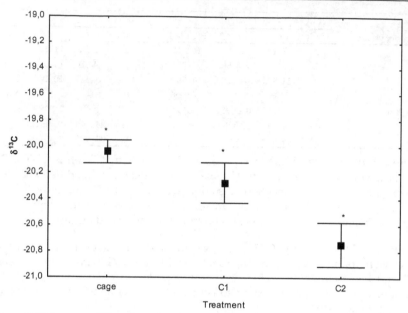

Fig. 5. Mean δ13C (mean ± SE) values for the three studied years of bivalve *Mytilus galloprovincialis* in the study sites: cage, control 1 (C1) and control 2 (C2). (PERMANOVA test, *p<0.05).

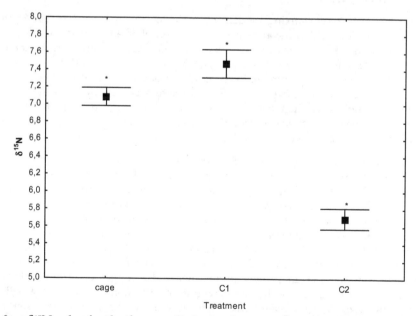

Fig. 6. Mean δ15N value for the three studied years (mean ± SE) of bivalve *Mytilus galloprovincialis* in the study sites: cage, control 1 (C1) and control 2 (C2) . (PERMANOVA test, *p<0.05).

Integrated Multitrophic Aquaculture: Filter Feeders Bivalves as Efficient Reducers of Wastes Derived from
Coastal Aquaculture Assessed with Stable Isotope Analyses

131

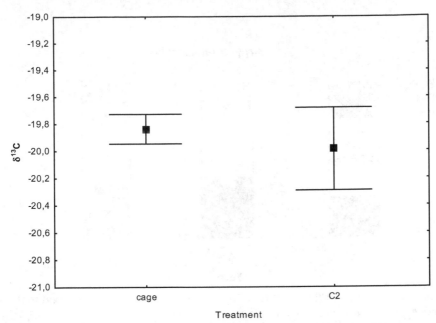

Fig. 7. Mean $\delta^{13}C$ (mean ± SE) values for the three studied years of bivalve *Chlamys varia* in the study sites: cage and control 2 (C2).

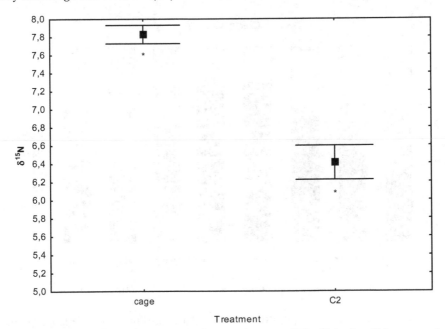

Fig. 8. Mean $\delta^{15}N$ value for the three studied years (mean ± SE) of bivalve *Chlamys varia* in the study sites: cage, and control 2 (C2). (PERMANOVA test, *$p < 0.05$).

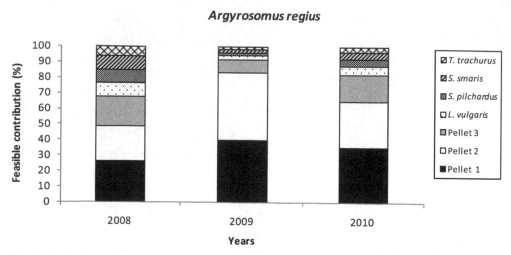

Fig. 9. Annual variation of the feasible contribution (%) of pellet and fresh food to the diet of cultured *Argyrosomus regius* based on Bayesian mixing models.

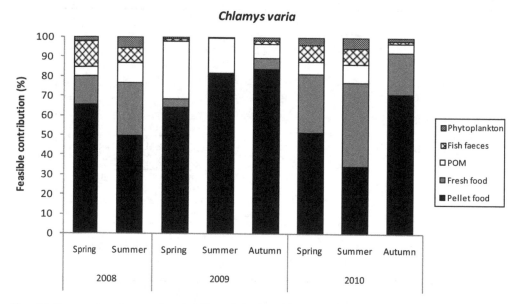

Fig. 10. Seasonal and annual variability of the feasible contribution of the main organic matter sources to the diet of *Chlamys varia* cultured in fish cages based on Bayesian mixing models.

Integrated Multitrophic Aquaculture: Filter Feeders Bivalves as Efficient Reducers of Wastes Derived from
Coastal Aquaculture Assessed with Stable Isotope Analyses

133

highest in spring 2009 (29.39 %). Especially remarkable is the low contribution of phytoplankton, with values from 0.06 % in summer 2009 to 5.27 % in summer 2008.

Similarly, pellet food was the main food source for *Mytilus galloprovincialis*, with a mean contribution of 57.95 %, followed by fresh food (21.46 %), fish faeces (8.90 %), POM (7.84 %) and phytoplankton (3.83 %). Contributions also varied seasonally and annually, but again aquaculture derived products remained the main food sources in the three years of study, with mean values varying from 82.79 % in spring 2009 to 97.7 % in spring 2008. Contrary to *Chlamys varia*, pellet food contribution was higher in spring and decreased in summer and autumn (Figure 11). Regarding the different commercial pellet food, similar to *Chlamys varia* pellet 1 was the main contributor, especially in spring 2008 were it represented up to 82 % of the total isotopic composition. *Sardina pilchardus* and *Spicara smaris* were the main sources among the fresh food supplied in fish cages. POM showed a lower contribution in the isotopic composition of *Mytilus galloprovincialis* than in *Chlamys varia*, with values ranging between 1.85 % in spring 2008 to 14.79 % in spring 2009. Contribution of phytoplankton was slightly higher than in *Chlamys varia*, with values between 0.87 % in spring 2008 to nearly 17.50 % in autumn 2009.

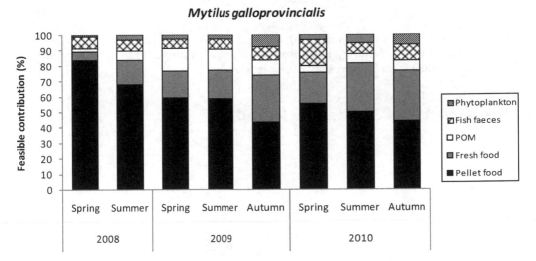

Fig. 11. Seasonal and annual variability of the feasible contribution of the main organic matter sources to the diet of *Mytilus galloprovincialis* cultured in fish cages calculated by Bayesian mixing models.

The mean feasible contribution of organic matter sources in sedimentivorous *Holothuria forskalii* was also analysed to study whereas there was an influence of organic enrichment in the benthic population (Figure 12). According to Bayesian mixing model, POM was the main food source, with a mean contribution of 48.56 % in the years of study. Contribution of aquaculture derived products as pellet and fresh food, and fish faeces was remarkable, with a 28.11 %. Bayesian mixing model was also applied to POM to elucidate its composition, with nearly the 50% of its isotopic signature coming from phytoplankton (Figure 13). The contribution of aquaculture derived products was also remarkable, with nearly 30 %.

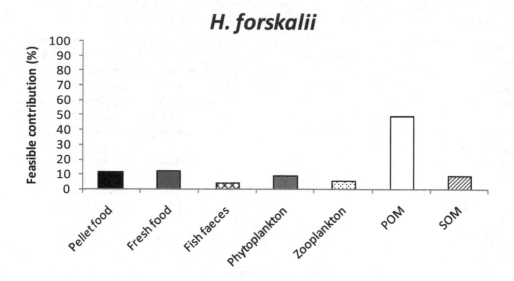

Fig. 12. Mean feasible contribution during the three years of study, of the main organic matter sources to the *Holothuria forskalii* isotopic signature, collected at the sea floor under fish cages and calculated by Bayesian mixing models.

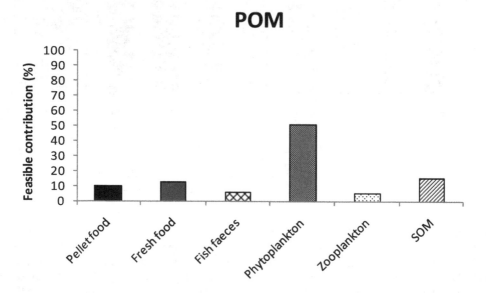

Fig. 13. Mean feasible contribution during the three years of study, of the main organic matter sources to the POM isotopic signatures collected at the sea floor, under fish cages, calculated by Bayesian mixing models.

Integrated Multitrophic Aquaculture: Filter Feeders Bivalves as Efficient Reducers of Wastes Derived from
Coastal Aquaculture Assessed with Stable Isotope Analyses

135

4. Discussion

The study is based on an approach of integrated multitrophic aquaculture to minimize the impact of aquaculture activities to the marine food web and benthic communities through stable isotope signatures of cultured *Argyrosomus regius*, filter feeder bivalves, aquaculture derived products, particulate and sedimentary organic matter at the proximity of cages. The results demonstrate the efficiency of integrated multitrophic aquaculture as valuable methods for the reduction of impacts of aquaculture activities at the coastal zone, especially so at oligotrophic waters such as the Balearic Islands.

Filter feeders bivalves (*Mytilus galloprovincialis* and *Chlamys varia*) are both enriched in ^{13}C and ^{15}N at the cage treatment when compared to the bivalves studied at the two control sites. This clearly demonstrates that both bivalves are assimilating the aquaculture derived wastes in an efficient way precluding that future aquaculture installations should incorporate species with different feeding strategies (fish, filter feeders) in order to guarantee a good environmental status by improving water quality at fish farms.

4.1 Food webs at cage and control treatments

The food web quantified by stable isotope signatures at each treatment (cage and two control sites) depicted different matter fluxes. ^{15}N isotopic signature can be used as a viable indicator to place trophic levels in a food web since nitrogen isotope 15 increases amongst trophic levels; therefore organisms higher in the trophic web will present higher δ^{15}N values. Accordingly, at the cage treatment the fish *Argyrosomus regius* has higher mean δ^{15}N values than filter feeders bivalves, zooplankton, phytoplankton, *Holothuria forskalii*, POM and SOM. Variability in ^{15}N isotopic signatures of pellet food reflects the pellet composition, which consists in a mixture of vegetal and fish meals with different proportion; pellet 2 is ^{15}N enriched due to its composition with 78 % fish meals and 22 % vegetal meals, while pellet 1 is ^{15}N-depleted because it is mainly composed of vegetal meals, with 71%. Previous studies have demonstrated that aquaculture derived waste enters the food web altering the natural isotopic composition of organic matter sources at the base and the upper trophic levels (Vizzini & Mazzola, 2004) as it is the case for our results.

In the cage treatment filter feeders bivalves showed a nitrogen isotopic signature similar to that one for zooplankton, phytoplankton, fresh food, *Argyrosomus regius* faeces and *Holothuria forskalii*. This similarity demonstrates that filter feeders are working efficiently by getting their nitrogen isotopic signature from the filtered substances. Phytoplankton, *Holothuria forskalii*, POM and SOM are carbon enriched compared to the other samples.

Figure 3 and 4 follow the same structure as Figure 2. While in cage treatment, POM is carbon enriched compared to SOM, in control 1, it is the other way round, SOM has higher values of δ^{13}C than POM. Nevertheless, filter feeders in control site 1 have nitrogen isotopic signatures higher than POM and SOM, which are in the lower part of the food chain, and isotopic signatures similar to those for zooplankton and phytoplankton. In Figure 3 we can observe how POM has δ^{13}C values much lower than the POM sampled in the cage site and control 2 site. We can also observe that in control site 2 all ^{15}N isotopic signature values for all samples are slightly lower than values found for samples in both cage site and control site 1. In control site 2 we find cleaner waters which are purified as well as by filter feeders by aquatic plants such as *Posidonia oceanica (Linnaeus) Delile 1813* which are present in smaller amounts or not even present in the cage site and control site 1. Aquatic plants give up dissolved nitrogen in waters, meaning that there are smaller amounts in the aquatic mean available for filter feeders, phytoplankton and zooplankton, resulting in a decrease of the isotopic signature for these compared to the isotopic signature values given in the other two study sites.

Regarding the filter feeder bivalves, *Mytilus galloprovincialis* has lower $\delta^{15}N$ and $\delta^{13}C$ values than *Chlamys varia* (Figures 5, 6, 7 and 8). Difference in $\delta^{13}C$ between species are smaller than differences in $\delta^{15}N$, the reason being that nitrogen is more stable than carbon. Nitrogen is used in amino acids and lipids while carbon is used in carbohydrates which are constantly being used up. Nitrogen therefore has a longer persistence in tissues than carbon, showing up with a higher isotopic signature. Moreover, nitrogen-rich fish waste affects $\delta^{15}N$ values, while $\delta^{13}C$ shows less variation as already stated by previous studies (Vizzini & Mazzola, 2004).

For all study sites, SOM always gives lower $\delta^{15}N$ values and higher $\delta^{13}C$ values in contrast to the other samples. Low values of SOM might be attributed to the fact that only the fine fraction of the sediment (<125 mm) was analysed for stable isotope determination. Another explanation could be a larger filtering capacity and high waste assimilation by filter feeders therefore less organic matter will be available for incorporation at SOM. Other studies (Sarà et al., 2004, 2006) have addressed effects of farming wastes to POM and SOM, but no inclusion of all organic sources and consumers have been considered. Our study improves knowledge on fish farms effects by integrating filter feeders bivalves.

The enrichment in ^{15}N exhibited by *Holothuria forskalii* in the cage site is derived by the incorporation of part of aquaculture wastes that sink to the seafloor, as it has been suggested by the Bayesian mixing model, and observed by other authors, as Dolenec et al. (2007). Even though it has been demonstrated that filter feeders are efficient in the assimilation of part of these products, there is still a fraction which accumulates in the sediment affecting the $\delta^{15}N$ signature of the sedimentivorous feeders. This is also reflected in control 1, which is ^{15}N enriched relative to control 2, where ^{15}N enrichment is probably due to the effect that the presence of cleaner waters and aquatic plant communities might have on sedimentivorous species.

Higher $\delta^{13}C$ values for *Chlamys varia* and *Mytilus galloprovincialis* in the cage site compared to the reference sites are probably linked to higher water residence time in the inner bay where the fish farm is deployed, while control sites are at open areas. Therefore, phytoplankton blooms at cage are more probable resulting in an increase of the $\delta^{13}C$ isotopic signature for both filter feeders.

4.2 Partial wastes contributions

The existing annual variability of feasible contribution of fresh food and pellet food to *Argyrosomus regius*'s diet (Figure 9) most probably depends on the nutritional regime that fin fish have been fed on; that is if they have been given more or less fresh food in contrast to pellet food. For the three years of study, the feasible contribution of pellet food compared to fresh food is considerable higher. During 2008, *Argyrosomus regius* shows a much higher feasible contribution of fresh food than during the other two years. The reason being is attributed merely to distinctions in the feeding mode and not due to a change in the intake nutrient strategy of *Argyrosomus regius*.

Integrated Multitrophic Aquaculture: Filter Feeders Bivalves as Efficient Reducers of Wastes Derived from
Coastal Aquaculture Assessed with Stable Isotope Analyses

137

Filter feeders bivalves in our study, *Mytilus galloprovincialis* and *Chamys varia*, show both seasonal and annual variability (Figure 10 and 11). During autumn 2009 and 2010 *Chlamys varia* has a higher feasible contribution due to pellet food (Figure 7) but it filter less pellet food during summer (2008 and 2010), with the exception of summer 2009. *Chlamys varia* increases the feasible contribution due to phytoplankton during spring and summer, a reason might be that during this time of the year phytoplankton blooms increase their presence increasing at the same time their contribution to filtered matter by *Chlamys varia*. The feasible contribution of phytoplankton, fish faeces and POM is very small and rather similar for *Chlamys varia* during autumn 2009 and 2010. During autumn there is a lower phytoplankton activity and less particulate organic matter and therefore the contribution of these will be lower. Therefore, the seasonal variability of *Chlamys varia* might indirectly indicate the seasonal variability of phytoplankton, POM and faeces.

Filtering capacity and assimilation of the different components of the food web by filter feeders bivalves present a remarkable pattern as observed from the mixing models calculated at *Mytilus galloprovincialis* and *Chlamys varia* (Figure 10 and 11). For both species, the partial contribution are complementary, in the sense that when one specie increases the feasible contribution due to pellet food, the other specie decreases the contribution due to this type of food and vice versa. This could be interpreted as a competitive strategy between both species, possibly due to being in the same bags limited by water income and nutrient availability. Another feasible explanation is that their diets differ seasonally, and meanwhile one is relying on phytoplankton, the other can be assimilating more aquaculture derived wastes and vice versa.

The influence of excess feed in the isotopic signal of the particulate organic matter (POM) of the water column in the cages is indicative of the impact on the marine ecosystem. Similarly, the high contribution of the particulate organic matter and debris supply to the isotopic signal of *Holothuria forskalii* reinforce the impact on the benthos. Thus, it would be interesting in future studies to take into account, as components of the food web, sedimentivores (which may further reduce the organic load in the sediment), opportunistic fish species (which may alter the carbon fluxes and benthic and pelagic nitrogen) and the microbial loop (which rejoins the nutrients back to the food web, assuming a renewal of the carbon and nitrogen present).

5. Acknowledgements

This work was financially supported by the JACUMAR project (Ministerio de Medio Ambiente y Medio Rural y Marino). The authors thank several people involved in samples processing such us F. Fuster, S. Sardu and M. Ceglia. Special thanks for collaboration in stable isotope analyses offered by the SCTI (Scientific-Technical Services) from the Balearic Island's University and B. Martorell. We appreciate the collaboration of the staff members of the marine protected area of S'Arenal.

6. References

Bergamino, L.; Lercari, D. & Defeo, O. (2011). Food web structure of sandy beaches: Temporal and spatial variation using stable isotope analysis. *Estuarine, Coastal and Shelf Science*, Vol. 91, No. 4, (March 2011), pp. 536-543, 02727714

Brown, J. R.; Gowen, R. J. & McLusky, D. S. (1987). The effect of salmon farming on the benthos of a Scottish sea loch. *Journal of Experimental Marine Biology & Ecology*, Vol. 109, No. 1, (July 1987), pp. 39-51, 00220981

Carabel, S.; Godínez- Dominguez, E.; Verísimo, P.; Fernández, L. & Freire, J. (2006). An assessment of sample processing methods for stable isotope analyses of marine food webs. *Journal of Experimental Marine Biology and Ecology*, Vol. 336, No. 2 (September 2006), pp. 254-61

Cheshuk, B.W.; Purser, G.J. & Quintana, R. (2003). Integrated open-water mussel (*Mytilus planulatus*) and Atlantic salmon (*Salmo salar*) culture in Tasmania, Australia. *Aquaculture*, Vol. 218, No. 1-4, (March 2003), pp. 257-378, 00448486

DeNiro, M. J. & Samuel, E. (1978). Influence of diet on the distribution of carbon isotopes in animals. *Geochimica et Cosmochimica Acta*, Vol. 42, No. 5 (May 1978), pp. 495-506

Dolenec, T.; Lojen, S.; Kniewald, G.; Dolenec, M. & Rogan, N. (2007). Nitrogen stable isotope composition as a tracer of fish farming in invertebrates *Aplysina aerophoba, Balanus perforatus* and *Anemonia sulcata* in central Adriatic. *Aquaculture*, Vol. 262, No. 2-4, (February 2007), pp. 237-249, 00448486

FAO Yearbook. 2008. *Fishery and Aquaculture Statistics*. FAO, ISBN 978-9250066981, Rome, Italy

FAO. 2010. *The state of world Fisheries and Aquaculture*. FAO, ISBN 978-92-5-306675-9, Rome, Italy

Holmer, M.; Argyrou, M.; Dalsgaard, T.; Danovaro, R.; Diaz-Almela, E.; Duarte, C.; Frederiksen, M.; Grau, A.; Karakassis, I.; Marbá, N.; Mirto, S.; Pérez, M.; Pusceddu, A. & Tsapakis, M. (2008). Effects of fish farm waste on *Posidonia oceanica* meadows: Synthesis and provision of monitoring and management tools. *Marine Pollution Bulletin*, Vol. 56, No. 9, (September 2008), pp. 1618-1629, 0025326X

Karakassis, I. & Hatziyanni, E. (2000). Benthic disturbance due to fish farming analyzed under different levels of taxonomic resolution. *Marine Ecology Progress Series*, Vol. 203, (September 2000), pp. 247-253, 01718630

Mazzola, A. & Sará, G. (2001).The effect of fish farming organic waste on food availability for bivalve molluscs (Gaeta Gulf, Central Tyrrhenian, Med): Stable carbon isotopic analysis. *Aquaculture*, Vol. 192, No. 2-4 (January 2001), pp. 361-79, 00448486

McCutchan, J.H.; Lewis, W. M.; Kendall, C. & McGrath, C. C. (2003). Variation in trophic shift for stable isotope ratios of carbon, nitrogen, and sulfur. *Oikos*, Vol. 102, No. 2 (August 2003), pp. 378-90, 00301299

Mente, E.; Pierce, G. J.; Santos, M. B. & Neofitou, C. (2006). Effect of feed and feeding in the culture of salmonids on the marine aquatic environment: A synthesis for European aquaculture. *Aquaculture International*, Vol. 14, No. 5 (October 2006), 499-522, 09676120

Michener, R. H & Lajtha, K. 2007. Stable isotopes in ecology and environmental science. Wiley-Blackwell, ISBN 978-1-4051-2680-9.

Mirto, S.; Bianchelli, S.; Gambi, C.; Krzelj, M. K.; Pusceddu, A.; Mariaspina, S.; Holmer, M. & Danovaro, R. (2010). Fish-farm impact on metazoan meiofauna in the

Integrated Multitrophic Aquaculture: Filter Feeders Bivalves as Efficient Reducers of Wastes Derived from
Coastal Aquaculture Assessed with Stable Isotope Analyses

139

Mediterranean Sea: analysis of regional vs. habitat effects. *Marine Environmental Research*, Vol. 69, No. 1, (February 2010), pp. 38-47, 01411136

Phillips, D. L. & Gregg, J. W. (2001). Uncertainty in source partitioning using stable isotopes. *Oecologia*, Vol. 127, No.2 (July 2001), pp.171-179

Reid, G. K.; Liutkus, M.; Bennett, A.; Robinson, S. M. C. & MacDonald, B. (2010). Absorption efficiency of blue mussels (*Mytilus edulis* and *M. trossulus*) feeding on Atlantic salmon (*Salmo salar*) feed and fecal particulates: Implications for integrated multi-trophic aquaculture. *Aquaculture*, Vol. 299, No. 1-4, (February 2010), pp. 165-169

Sarà, G.; Scilipoti, D.; Mazzola, A. & Modica, A. (2004). Effects of fishing farming waste to sedimentary and particulate organic matter in a southern Mediterranean area (Gulf of Castellammare, Sicily): a multiple stable isotope study (δ^{15}N and δ^{13}C). *Aquaculture*, Vol. 234, No. 1-4, (May 2004), pp. 199-213,

Sarà, G.; Scilipoti, D.; Milazzo, D. & Modica, A; (2006). Use of stable isotopes to investigate dispersal of waste from fish farms as a function of hydrodynamics. *Marine Ecology Progress Series*, Vol. 313, (May 2006), pp. 261-270, 01718630

Shpigel, M. & Blaylock, R. A.(1991). The Pacific oyster, *Crassostrea gigas*, as a biological filter for a marine fish aquaculture pond. *Aquaculture*, Vol. 92, No. 2-3, (1991), pp. 187-197, 00448486

Skalli, A.; Hidalgo, M. C.; Abellán, E.; Arizcun, M. & Cardenete, G. (2004). Effects of the dietary protein/lipid ratio on growth and nutrient utilization in common dentex (*Dentex dentex* L.) at different growth stages. *Aquaculture*, Vol. 235, No. 1-4, (June 2004), pp. 1-11, 00448486

Soto, D.; Aguilar-Manjarrez, J.; Bermúdez, J.; Brugère, C.; Angel, D.; Bailey, C.; Black, K.; Edwards, P.; Costa-Pierce, B.; Chopin, T.; Deudero, S.; Freeman, S.; Hambrey, J.; Hishamunda, N.; Knowler, D.; Silvert, W.; Marba, N.; Mathe, S.; Norambuena, R.; Simard, F.; Tett, P.; Troell, M. & Wainberg, A. (2008). Applying an ecosystem-based approach to aquaculture: principles, scales and some management measures. In: FAO Fisheries and Aquaculture Proceedings 11, FAO

Stirling, H. & Okumus, I. (1995). Growth and production of mussels (*Mytilus edulis*) suspended at salmon cages and shellfish farms in two Scottish sea lochs. *Aquaculture*, Vol. 134, (July 2005), pp. 193-210

Taylor, B. E.; Jamieson, G. & Carefoot, T. H. (1992). Mussel Culture in British-Columbia - the Influence of Salmon Farms on Growth of *Mytilus edulis*. *Aquaculture*, Vol. 108, No. 1-2 (November 1992), pp. 51-66.

Troell, M.; Halling, C.; Neori, A.; Chopin, T.; Buschmann, A. H.; Kautsky, N. & Yarish, C. (2003). Integrated mariculture: asking the right questions. *Aquaculture*, Vol. 226, No. 1-4, (October 2003), pp. 69–90, 00448486

Vizzini, S. & Mazzola, A. (2004). Stable isotope evidence for the environmental impact of a land-based fish farm in the western Mediterranean. *Marine Pollution Bulletin*, Vol. 49, No. 1-2, (July 2004), pp. 61-70, 0025326X

Weston, D. (1991). Quantitative examination of macrobenthic community changes along an organic enrichment gradient. *Marine Ecology Progress Series*, Vol.61, (March 1990), pp.233-244

Yokoyama, H.; Abo, K. & Ishihi, Y. (2006). Quantifying aquaculture-derived organic matter in the sediment in and around a coastal fish farm using stable carbon and nitrogen isotopes ratio. *Aquaculture*, Vol. 254, No. 1-4, (April 2006), pp. 411-425, 00448486

8

Efficacy of Pilot-Scale Wastewater Treatment upon a Commercial Recirculating Aquaculture Facility Effluent

Simonel Sandu[1], Brian Brazil[2,3] and Eric Hallerman[1]
[1]*Virginia Polytechnic Institute and State University*
[2]*U.S. Department of Agriculture – Agricultural Research Service*
[3]*Geosyntec Consultants, N.W., Kennesaw, GA*
USA

1. Introduction

Recirculating aquaculture systems have been developed to produce high-value species for year-round supply of markets, to free production from site-related constraints, to minimize environmental impacts of aquaculture, and to make efficient use of limited high-quality water supplies. Effective treatment and reuse of aquaculture effluent has been demonstrated at an experimental scale. Blue Ridge Aquaculture (BRA, Martinsville, Virginia, USA) is a commercial venture producing 1360 metric tons of hybrid tilapia (*Oreochromis* sp.) per year in recirculating aquaculture systems. To our knowledge, BRA is the largest recirculating aquaculture enterprise under one roof in existence. Increased production at BRA is constrained by the availability of high-quality influent water. Meanwhile, BRA discharges an estimated 2290 m^3 of wastewater per day to the municipal sewer system, equivalent to an average of 1 m^3 discharged per 3.0 kg feed. Importantly, discharge of this effluent also loses heat energy, as water temperature is maintained at about 28 - 30°C to optimize tilapia growth. Concerned about the reliability and costs of their wells and the city water system, BRA seeks a technical solution. Developing a wastewater treatment system that recovers and reuses the water presently discharged could minimize these problems. The discharge issues faced by BRA typify intensive aquaculture, and evaluation of possible treatment strategies would have general interest to the aquaculture sector.

The BRA waste stream was characterized over eight-hour sampling periods during 12 different days. The results indicated that solids, chemical oxygen demand (COD), and nitrate were the most significant waste components by concentration and weight. All these forms of pollutants were targeted by unit processes in a treatment train. Following an ozone treatability study (Sandu, 2004), a pilot-scale wastewater treatment station was built in order to initiate, characterize and optimize the operation of this more complex treatment strategy. The ultimate goal was to obtain an effluent clean enough to be reused in the recirculating systems. The need to eliminate settable solids, colloids, dissolved organic substances and nitrogenous compounds led to selection of a sequential treatment process employing physical, biological, chemical, and again, biological steps. The performance of the pilot-scale wastewater treatment train with regard to solids and carbonaceous compounds was reported by Sandu et al. (2008). Here, we: (1) evaluate the effectiveness of a pilot wastewater treatment train with regard to alkalinity, pH, hardness, and nitrogenous compounds (TAN, NO_2^--N, NO_3^--N and TKN), (2) the effect of nitrate feed rate on denitrification, and (3) examine the economic feasibility of treating and reusing this aquaculture wastewater. All abbreviations used in this chapter are presented in Table 1.

BRA	Blue Ridge Aquaculture
BOD	biochemical oxygen demand (mg/l)
$cBOD_5$	carbonaceous biochemical oxygen demand (mg/l)
CF	chemical flocculation
COD	chemical oxygen demand (mg/l)
DR	denitrification reactor
NO_2^--N	nitrite-nitrogen (mg/l)
NO_3^--N	nitrate-nitrogen (mg/l)
NO_x^--N	oxidized nitrogen species (mg/l)
OR	ozone reactor
RI	raw (untreated) influent
SB	sedimentation basin
TAN	total ammonia nitrogen (mg/l)
TF	trickling filter
TKN	total Kjeldall nitrogen (mg/l)
TSS	total suspended solids (mg/l)
VSS	volatile suspended solids (mg/l)
Y_b	biological (anoxic) yield (g microbial biomass produced/g substrate used)
Y_{NO3^--N}	biological yield for NO_3^--N (g biomass VSS/g NO_3^--N)
Y_{VSS}	biological yield for VSS (g biomass VSS/g dissolved COD)
Y_{COD}	biological yield for COD (g biomass COD/g dissolved COD)

Table 1. Nomenclature used in this chapter.

2. Methods

2.1 Treatment train and experimental treatments

The wastewater treatment train (Figure 1) included storage tanks, a primary sedimentation basin, mechanical filtration using a microscreen drum filter, denitrification using a fluidized bed biological reactor with methanol added to provide carbon and energy for cellular growth, ozonation and foam fractionation in a bubble-contact ozone reactor, dissolved ozone quenching in an air-bubble stripping chamber, aerobic biological treatment using a trickling filter, and jar test-scale chemical flocculation, followed by sand filtration. A detailed description of the pilot plant, its operation and analytical techniques may be found in Sandu (2004) or Sandu et al. (2008).

Our evaluation of pilot plant effectiveness consisted of four different experimental treatments (Table 2), differing by use of 6 or 4 lpm flow and recycling rates, ozone doses between 36.6 – 82.5 mg O_3/l water, and 6- or 9-minute ozonation times.

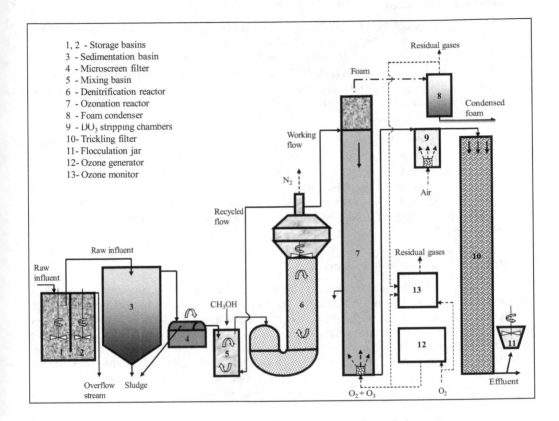

Fig. 1. Schematic diagram of pilot-scale wastewater treatment train at Blue Ridge Aquaculture. Details are not drawn to scale.

Treatment	Water flow lpm	Gas flow[1] lpm	O_3 conc. mg O_3/l gas	O_3 dose mg O_3/l water	Oz. time min	Recircul. DR[2] lpm (%)	TF used[3] %
1	6		22	36.6	6	4 (40%)	100
2		10	33	55.0			
3	4		33	82.5	9	6 (60%)	50
4			22	55.0			

[1]Flow of the O_2/O_3 mixture.
[2]Recirculation rate in the denitrification reactor.
[3]Proportion of cross section of trickling filter used.

Table 2. Controlled parameter conditions applied to the pilot station for the four experimental treatments.

2.2 Water quality assays

Water samples collected for this study were passed through a 0.45 μm filter and tested for total ammonia nitrogen (TAN), nitrate-nitrogen (NO_3^--N) and nitrite-nitrogen (NO_2^--N) content using a spectrophotometer (Hach DR 2400, Loveland, CO). Temperature and pH were determined by using a pH/mV/°C meter (Oakton® Acorn Meter Kit model pH 6, Vernon Hills, Illinois). Alkalinity and hardness were measured using Hach Permachem® Reagent methods. Total Kjeldall nitrogen (TKN) was determined by using macro-Kjeldall Standard Method 4500-N_{org} B, (American Public Health Association [APHA] et al., 1998).

For statistical analyses, 95% simultaneous confidence intervals were determined to compare means of water quality parameters after passage through each of the five main units of the treatment train. We used a one-way ANOVA with the four treatments as the factor in the model; the values of the parameters in the influent water were added as covariates for ANOVAs performed for the sedimentation basin and the denitrification reactor. For the units following the denitrification reactor, covariates were not used because they did not improve the model.

2.3 Biological yields determination

The biological (anoxic) yield, Y_b (i.e., the amount of microbial biomass formed per unit of substrate used), was determined using a bench-scale batch reactor for both methanol (as COD) and NO_3^--N substrates. The biomass produced was quantified as volatile suspended solids (VSS) and COD. A hermetically closed 5-L vessel was initially filled with water seeded with biofilm sheared from a sample of coated sand from the denitrification reactor. Potassium nitrate and methanol were added in amounts accounting for initial concentrations of NO_3^--N and methanol (as dissolved COD) of 207 and 1666 mg/l, respectively. Methanol was well in excess of stoichiometric requirements so that it would not be a limiting factor. Continuous stirring and a constant temperature of 28±1°C were maintained. Pure nitrogen was injected for about 5 minutes to strip dissolved oxygen and to replace the air from the space above the fluid. Gas exiting at the top was collected and directed by a hose to a water bath, sealing the space. Any additional nitrogen gas produced by denitrification followed the same path before entering the atmosphere. The operation of purging nitrogen was repeated after any sample collection (i.e., at the beginning and at 12-hour intervals through 48 hours). Total suspended solids (TSS), volatile suspended solids (VSS), dissolved and total chemical oxygen demand (COD) were measured according to Standard Methods 2540D, 2540E, and 5220C, respectively (APHA et al., 1998). NO_3^--N was determined spectrophotometrically (Hach DR 2400, Loveland, CO). Filters of 1.5-μm pore diameter were used for collecting TSS, and for obtaining samples for dissolved COD determinations. The equations used to determine Y_b values (Grady et al., 1999) were:

$$Y_{VSS} = \frac{\Delta VSS}{\Delta\ dissolved COD} \quad \left(\frac{g}{g}\right) \tag{1}$$

and:

$$Y_{NO_3^- - N} = \frac{\Delta VSS}{\Delta\ NO_3^- - N} \quad \left(\frac{g}{g}\right) \tag{2}$$

Y_{COD} was obtained by the equation:

$$Y_{COD} = 1.42 \times Y_{VSS} \left(\frac{g}{g} \right) \tag{3}$$

Additionally, Y_{NO3}-N was determined from data collected from the denitrification reactor using Equation 2.

3. Results and discussion

3.1 BRA waste stream characterization

Characteristics of BRA aquaculture effluent regarding alkalinity, pH, hardness, and nitrogenous compounds are presented in Table 3.

Parameter	Average[1]	Minimum	Maximum
Alkalinity (mg/l)	118	76	141
Hardness (mg/l)	150	139	170
pH	7.21	6.95	7.62
TKN (mg/l)	31.5	5.11	47.36
TAN (mg/l)	2.55	1.72	4.11
NO_3^--N (mg/l)	42.98	6.8	68.8
NO_2^--N (mg/l)	0.91	0.45	1.93
TAN (mg/l)	2.55	1.72	4.11

[1] Average values represent non-flow-weighted averages (12% for samples from 2:00 p.m. and 44% each for the other times).

Table 3. Waste stream characteristics for BRA effluent collected on different days at 6:00 a.m., 2:00 p.m. and 10:00 p.m. each day.

3.2 Alkalinity

Alkalinity is managed at BRA by addition of industrial-grade sodium bicarbonate, $NaHCO_3$, which buffers pH and replaces alkalinity lost by nitrification and water exchange. Data from operation of the pilot station (Table 4) showed that alkalinity lost by water exchange was recovered and, further, that treated effluent was enriched by up to one-third of the initial amount of alkalinity. Alkalinity was generated in the denitrification reactor, increasing by approximately 40% in all experimental treatments. Despite the significant increase in percentage terms, the net production represented only about 2 mg alkalinity per mg of NO_3^--N reduced. Because the stoichiometry of denitrification reactions would lead us to expect a ratio of about 3.6, we infer that some NO_3^--N must have been transformed into ammonia by assimilative reduction and used in cell synthesis when ammonia was lacking (Grady et al. 1999). The inference was supported by the observation that the wastewater treated was low in TAN, with a ratio of TAN: NO_3^--N of about 1:20.

The stream entering the ozonation reactor had an alkalinity of 175.5-187.5 mg/l. During ozonation, between 6-12% of alkalinity was lost. Loss of alkalinity could be due to the

Treatment	SB mg/l	DR mg/l (%)	OR mg/l (%)	TF mg/l (%)	Total increase[2] (%)
1	114.5	192.1 (+40.4)[a]	175.5 (-6.6)[a]	163.6 (-6.8)[a]	30.0[a]
2			179.9 (-6.4)[a]	169.3 (-5.9)[a]	32.4[a]
3	121.7	201.8 (+39.7)[a]	187.5 (-7.1)[a]	179.5 (-4.2)[a]	32.2[a]
4			177.8 (-11.9)[a]	164.3 (-7.5)[a]	26.0[a]

[1]Abbreviations: SB = sedimentation basin, DR = denitrification reactor, OR = ozonation reactor, and TF = trickling filter.
[2]Alkalinity increase between sedimentation basin and trickling filter effluent concentrations.

Table 4. Dynamics of alkalinity through the pilot plant for all treatments (unit outlet values) [1]. Means in a column with the same superscript are not significantly different ($p>0.05$).

scavenging effect of its ions on ozone. HCO_3^- and CO_3^{2-} ions compete in wastewater with organic matter for reaction with the OH^o radical, and high alkalinity can impair the reaction of ozone with targeted organics (Wang & Pai 2001); alkalinity depletion increases with ozone dose due to increased probability of OH^o radical formation due to faster organics removal. The relatively low removal of alkalinity that we observed may have been limited by the pH of slightly over 8.0, not high enough for carbonate ion formation, and with ozone being less reactive with bicarbonate ion. Buxton et al. (1988) found the reaction rate constants for reaction of hydroxyl ion to be 39×10^7 l/mol-s for CO_3^{-2} ions and 0.85×10^7 l/mol-s for HCO_3.

More alkalinity was removed in the tricking filter, probably due to nitrification. This observation was not surprising, because stoichiometrically 1g of TAN can destroy 8.62 g of HCO_3^- during oxidation to NO_3-N (Grady et al., 1999). However, because TAN concentration was relatively low and because the nitrification was not complete (i.e., NO_2^--N was produced in the biofilter), the final effluent still had 26.0 – 32.6% more alkalinity than the stream entering the wastewater treatment train. Hence, reuse of this treated effluent could result in savings regarding supplemental alkalinity addition to the aquaculture system.

3.3 pH

The pH of the aquaculture effluent was neutral or slightly basic (Table 5), close to that of water in the fish production tanks. Neutral pH in wastewater can be due to the presence of inorganic salts (Millamena 1992) or to the heterogeneous composition of its organic matter (Medley & Stover 1983); BRA effluent exhibited both of these characteristics. After entering the treatment train, pH increased slightly in the storage tanks and settling basin, and then increased more significantly during denitrification, reaching values between 8.22 and 8.26 in denitrification reactor effluent. The pH increase was probably due to intense biological activity in these units, especially in the denitrification reactor, where pH increase was promoted by alkalinity generation. During ozonation, pH decreased, probably because some alkalinity was lost to attack by ozone-derived radicals. Kirk et al. (1975) found that whether ozonation feed water is acidic or basic, the product water always shifts toward neutrality, and that the pH change is greater for higher-COD feedwaters. Further, Wang & Pai (2001) suggested that the greatest organics removal by ozonation is obtained at low pH,

Treatment	RI	SB	DR	OR	TF	CF
1	7.22	7.30[a]	8.26[a]	8.09[a]	8.32[a]	7.61[a]
2				7.89[a]	8.24[a]	7.63[a]
3	7.20	7.37[a]	8.22[a]	7.93[a]	8.32[a]	7.58[a]
4				7.89[a]	8.24[a]	7.58[a]

[1] Abbreviations: RI = raw influent, SB = sedimentation basin, DR = denitrification reactor, OR = ozonation reactor, TF = trickling filter, and CF = chemical flocculation.

Table 5. Dynamics of pH values through the treatment train[1] for all experimental treatments (unit outlet values). Means in a column with the same superscript are not significantly different ($p>0.05$).

suggesting that the relatively high-pH water in our study could be disadvantageous. pH increased again in the trickling filter, which was somewhat surprising considering that nitrification occurred in this filter and some alkalinity was lost. The effects of large amounts of organics in the influent and their compositional diversity may have overshadowed the effects of nitrification of relatively small amounts of ammonia. Additionally, CO_2 stripping could have contributed to pH increase.

Further decline of pH occurred during chemical flocculation, reaching final values of about 7.60, which is considered safe for fish production.

3.4 Hardness

Hardness enters the BRA systems with replacement spring water and with feed, and supports biomass development for both fish and for microorganisms in biofilters. Average hardness concentrations were between 148.4 and 151.5 mg/l as $CaCO_3$, characterizing BRA effluent as a medium-hard wastewater. Table 6 presents the dynamics of hardness as the water passed through the treatment station.

Treatment	SB	DR		OR		TF		Total decrease[2]
	mg/l	mg/l	(%)	mg/l	(%)	mg/l	(%)	(%)
1	151.5[a]	145.7	(3.8)[a]	139.7	(4.1)[a]	138.0	(1.2)[a]	8.9[a]
2				137.4	(5.7)[b]	135.9	(1.1)[a]	10.3[a]
3	148.4[a]	141.7	(4.5)[a]	130.6	(7.8)[ab]	129.5	(0.8)[a]	12.7[a]
4				134.6	(5.0)[ab]	132.6	(1.5)[a]	10.6[a]

[1] Abbreviations: SB = sedimentation basin, DR = denitrification reactor, OR = ozonation reactor, and TF = trickling filter.
[2] Hardness decrease between settling basin and trickling filter effluent concentrations.

Table 6. Dynamics of hardness through the pilot plant[1] for all experimental treatments (unit outlet values). Means in a column with the same superscript are not significantly different ($p>0.05$).

Small amounts of hardness were lost between the settling basin and denitrification reactor effluents, suggesting that it was removed by microscreen filtration after bonding with dissolved and colloidal organics, or was assimilated during bacterial growth in the denitrification reactor.

Hardness decreased further during ozonation, with the decrease proportional to the amount of ozone applied (e.g., 4.1% removal in Treatment 1 and 7.8% removal in Treatment 3, a statistically significant difference). Ozone-induced hardness removal was probably because increased carboxyl acid concentration due to ozonation led to greater magnesium and calcium association, resulting in precipitation of metal-humate complexes (Grasso & Weber 1988). Higher hardness removal at higher ozone doses was attributable to more destabilized carboxyl acids binding with a larger amount of hardness ion species. Noting that increasing hardness to 150 mg/L $CaCO_3$ in ozonated water improved removal of TSS (Rueter & Johnson 1995), BRA effluent exhibited favorable conditions for TSS removal. After being formed in the aqueous medium, TSS complexes likely were caught on bubble surfaces and buoyed to the top as foam, explaining the high TSS removal efficiency during ozonation (Sandu et al. 2008). Once in the foam, however, the association of hardness with TSS seems to have been attacked by ozone, which broke and dissolved TSS particles.

Small amounts of hardness were lost in the trickling filter, probably due to bacterial assimilation and bonding into solids.

Although hardness was not greatly reduced during treatment, it could become depleted below critical limits after repeated treatments, and hence may require periodic adjustment.

3.5 TKN

In raw influent, TKN ranged between 41.8-42.3 mg/l, mostly as organic nitrogen. Table 7 shows TKN dynamics through the entire treatment train. TKN was removed in approximately the same proportion as COD during sedimentation, indicating that nitrogen-containing organics were distributed similarly between solid and dissolved forms. A slight decrease in TKN was observed in the denitrification reactor, which could be attributed to ammonia consumption in this reactor.

Treatment	RI mg/l	SB mg/l	SB (%)	OR mg/l	OR (%)	CF mg/l	CF (%)	Total removal[2] (%)
1	41.8	16.9	(59.6)	9.2	(45.6)[a]	2.0	(78.7)[a]	95.2[a]
2				8.2	(51.5)[ab]	1.9	(76.8)[a]	95.5[a]
3	42.3	16.2	(61.7)	7.4	(54.3)[b]	1.7	(77.0)[a]	96.0[a]
4				8.2	(49.4)[ab]	1.9	(76.8)[a]	95.5[a]

[1]Abbreviations: RI = raw influent, SB = sedimentation basin, OR = ozonation reactor, and CF = chemical flocculation.
[2]Decline of TKN between raw influent and trickling filter effluent concentrations.

Table 7. Total Kjeldall nitrogen (TKN) dynamics through the pilot plant[1] for each treatment (unit outlet values). Means in a column with the same superscript are not significantly different ($p > 0.05$).

Ozonation removed up to 54% of influent TKN (at the highest ozone dose), which exceeded the proportion of COD removal. There was a statistically significant difference in TKN removal between Treatments 1 and 3, suggesting that TKN removal rate depended on ozone dose. Nitrogen-containing compounds are more prone to ozone-mediated destabilization than many other organics, facilitating bonding with opposite electrical charges (Razumovskii & Zaikov, 1984). In this case, N-containing compounds likely bonded directly to charges at the surface of gas bubbles (fractionation effect) or with polyvalent ions, and subsequently were removed with foam. Some of these molecules also were mineralized, which is reflected in the ammonia increase during ozonation. In turn, NO_3^--N rose slightly during ozonation as some ammonia was oxidized further due to favorable conditions in an alkaline environment and pH above 8 (Lin & Wu 1996). Our results showed a generally higher TKN removal than other studies. For example, Beltran et al. (2001) found a 26% TKN removal at ozone doses between 40-60 mg/l on domestic wastewater that had been treated biologically. Higher TKN removal in our study could be attributed to higher alkalinity in BRA wastewater and to different composition of organics in the two wastewaters.

In the trickling filter, TKN was reduced by 15-31%. The percent removal did not appear to depend on the ozone dose applied in the previous treatment step. Although an increase in ozone dose should promote TKN removal (Beltran et al. 2001), our finding differed, probably because a great part of TKN was in the form of ammonia after ozonation. In this circumstance, TKN removal efficiency was rather dependent on the nitrification performance of the trickling filter.

TKN removal by chemical flocculation ranged from 77-79%. Comparing the average of 1.7-2.0 mg/l for TKN after chemical flocculation to an average of 1.5-1.7 mg/l for TAN, it is clear that the organic component of TKN was almost entirely removed by the treatment train.

3.6 TAN

Table 8 shows TAN dynamics through the treatment train. The average influent TAN concentration ranged between 2.53 and 2.58 mg/l in all experimental treatments. These values were higher than the average of 2.06 mg/l in the recirculating aquaculture systems, with the increase likely due to bacterial activity in the storage tanks and sedimentation basin. Ammonia is utilized preferentially as a nitrogen source by heterotrophic bacteria (Grady et al. 1999), explaining the 48-50% reduction of TAN as the stream underwent denitrification.

During ozonation, TAN concentration rose higher than influent levels by a treatment average of 29-40%. These TAN concentrations were over twice those in the denitrification reactor effluent in Treatments 2 and 3. The increase of TAN concentration during ozonation exhibited a positive, linear relationship with ozone dose (slope = 0.012; r^2 = 0.93). The increase of TAN probably was due to amino acid and protein oxidation by ozone. Ammonia is a byproduct of these reactions, especially when they are complete (i.e., mineralization). The basic pH of the ozonation reactor influent appeared to promote partial oxidation of ammonia to nitrate, because NO_3^--N increased by more than expected from influent NO_2^--N oxidation. However, the oxidation reaction was insignificant and ammonia accumulation predominated. Rosenthal & Otte (1979) and Wang & Pai (2001) also reported partial oxidation of ammonia to NO_3^--N during ozonation under alkaline conditions, with TAN accumulating via oxidation of organic nitrogen.

Treatment	SB mg/l	DR mg/l (%)	OR mg/l (%)	OR % Increase[2]	TF mg/l (%)	Total[3] (%) removal
1	2.58	1.33 (-48.4)a	3.64 (+173.6)a	29.1	1.69 (-53.6)a	34.5a
2			3.92 (+194.7)a	34.2	1.59 (-59.4)a	38.4a
3	2.53	1.26 (-50.2)a	4.21 (+234.1)a	39.9	1.52 (-63.9)a	39.9a
4			4.00 (+217.5)a	36.8	1.63 (-59.3)a	35.6a

[1]Abbreviations: SB = sedimentation basin, TF = trickling filter, OR = ozonation reactor, and DR = denitrification reactor.
[2]Percent increase of TAN concentration after ozonation from the initial TAN concentration in the sedimentation basin.
[3]Percent decrease of TAN concentration after ozonation from the initial TAN concentration in the sedimentation basin.

Table 8. Total ammonia nitrogen (TAN) dynamics through the pilot plant[1] for all treatments (unit outlet values). Means in a column with the same superscript are not significantly different ($p>0.05$).

In the trickling filter, partial nitrification occurred, removing 54-64% of TAN along with organics. The organic loading of the trickling filter was estimated at 0.43 kg $cBOD_5/m^3$-d in treatments 1 and 2, and 0.65 kg $cBOD_5/m^3$-day in Treatments 3 and 4. At these loadings, conditions were not permissive for nitrifiers to grow and compete effectively with heterotrophs, which made the nitrification performance of the trickling filter surprisingly good. In comparison, Metcalf & Eddy (1979, as cited in Karnchanawong and Polprasert, 1990) obtained 75–85% TAN removal in a trickling filter at lower volumetric loadings of 0.10 to 0.16 kg BOD_5/m^3-day. Parker & Richards (1986) suggested a maximum threshold of 27 mg/l BOD_5 in order for any nitrification to occur in a trickling filter. Our results also showed more efficient TAN removal when the stream had less organics, as in Treatment 3, although we measured $cBOD_5$ instead of soluble BOD_5.

The final effluent had TAN treatment averages between 1.52 and 1.69 mg/l, which is generally undesirable in water used for exchange in aquaculture systems. However, were this treated water used for exchange, only 0.84% of daily TAN production would be reintroduced and the rotating biological contactors in the fish production systems would be able to remove these amounts (Sandu et al. 2008).

3.7 NO_2-N

Nitrite results from incomplete nitrification in the aquaculture systems' rotating biological contactors. The average concentration was between 0.92 and 0.96 mg/l NO_2^--N in BRA effluent. NO_2^--N concentration fluctuated through the treatment train (Table 9).

In the denitrification reactor, between 72-76% of influent NO_2^--N was reduced to nitrogen. This reduction suggests that the external carbon source was supplied in an amount sufficient to support the completion of denitrification (van Rijn & Rivera 1990) and that influent nitrite also was reduced in this process. Another mechanism for nitrite reduction could be its utilization as a source of nitrogen by heterotrophic organisms in the upper part of the biofilter due to the relatively low concentration of TAN in the stream. While we cannot conclude which of these factors drove it, we regard NO_2^--N reduction in the

Treatment	SB mg/l	DR mg/l (%)	OR mg/l (%)	TF mg/l	TF (%)[2]	Total removal[3] (%)
1	0.96	0.23 (76.0)	0.00 (24.0)	0.45	(+55.6)[a]	44.4[a]
2				0.54	(+52.9)[a]	47.1[a]
3	0.92	0.26 (71.7)	0.00 (28.3)	0.61	(+58.7)[a]	41.3[a]
4				0.51	(+63.8)[a]	36.2[a]

[1]Abbreviations: SB = sedimentation basin, TF = trickling filter, DR = denitrification reactor, and OR = ozonation reactor.
[2]Percentages for the trickling filter express the mass generated as a fraction of the treatment train influent concentrations.
[3]Total percent decrease of NO_2^--N concentration relative to initial TAN concentration in the sedimentation basin.

Table 9. Nitrite-nitrogen (NO_2^--N) dynamics through the pilot plant[1] for all treatments (unit outlet values). Means in a column with the same superscript are not significantly different ($p>0.05$).

denitrification reactor as a positive outcome from application of excess methanol. Although the excess methanol could impose a higher dissolved organics load on the ozone reactor, this outcome may be preferred to poorer removal of nitrogenous compounds.

The remaining NO_2^--N then was oxidized totally to NO_3^--N in the ozone reactor, regardless of the ozone dose applied. Rosenthal & Otte (1979) also found that even with light ozonation, NO_2^--N in aquaculture wastewaters can be oxidized efficiently to NO_3^--N.

Although the stream entering the trickling filter had little or no NO_2^--N, the effluent had an average of 0.45-0.61 mg/l NO_2^--N, varying among experimental treatments. This concentration represented 53-64% of the treatment train influent concentration. The generation of NO_2^--N in the trickling filter probably was due to incomplete nitrification of ammonia. One of the causes could be the lack of NO_2^--N itself as substrate in the influent, which did not support the growth of bacteria converting NO_2^--N to NO_3^--N (i.e., *Nitrobacter sp.*). Summerfelt (2003) suggested that lack of these species in nitrification biofilters can be a drawback of integrating an ozonation step in a treatment loop in a recirculating aquaculture system, although the decrease of nitrite levels is a substantial benefit. Another cause of nitrite generation could be suppressed growth of *Nitrobacter sp.* by faster-growing heterotrophs under conditions of abundant of organic material (Parker & Richards 1986). *Nitrobacter sp.* are the slowest-growing nitrifiers and are the first to be eliminated by heterotrophs in a biofilter when competing for space (Grady et al. 1999). Considering nitrite and organic concentrations coming into the trickling filter in our study, both mechanisms appear plausible explanations for nitrite accumulation.

The presence of nitrite is undesirable in waters used for exchange in recirculating aquaculture systems because of its toxicity to fish, although the concentrations in our final effluent did not present a threat to fish. Further, the rotating biological contactors in the BRA fish production systems would be able to remove the amounts of NO_2^--N returned with exchange water.

3.8 NO₃-N

Nitrate was the most abundant nitrogenous waster in BRA effluent, resulting from nitrification and accumulation in the system. Treatment averages were between 42.8 and 43.2 mg/l NO_3^--N in the influent (Table 10), although large diurnal variations were observed due to different representations of wastewaters from greenhouses and grow-out systems within the overall BRA operation. On average, the denitrification reactor removed 96-97% of NO_3^--N among experimental treatments, suggesting that the biofilter adapted rapidly to nitrate fluctuations. This observation agrees with Jeris & Owens' (1975) suggestion that under conditions of nitrate variation, it is sufficient to supply the right amount of carbon source at any time in order to obtain satisfactory denitrification. Nitrate removal performance appeared to be independent of the recycled stream fraction among different treatments.

Treatment	SB	DR		TF	Total removal[2]
	mg/l	mg/l	(%)	mg/l	(%)
1	43.2	1.8	(95.8)[a]	2.6[a]	94.0
2				2.7[a]	93.8
3	42.8	1.3	(97.0)[a]	2.0[a]	95.4
4				2.4[a]	94.4

[1]Abbreviations: SB = sedimentation basin, TF = trickling filter, and DR = denitrification reactor.
[2]Total percent decrease of NO_3^--N concentration relative to initial TAN concentration in the sedimentation basin.

Table 10. Nitrate-nitrogen (NO_3^--N) dynamics through the pilot plant[1] for all treatments (unit outlet values). Means in a column with the same superscript are not significantly different ($p>0.05$).

The nitrate removal efficiency of the treatment train overall was slightly lower than that of the denitrification reactor unit, as NO_3^--N was produced during ozonation and by nitrification in the trickling filter. However, the final concentration was generally less than 3 mg/l, posing no issue for reuse of the recovered wastewater for fish production.

3.9 Cell yields and the effect of nitrate fed on denitrification

From tests on the batch reactor, cell yield, Y_{NO3}-N, was 0.69 g VSS cells produced/g NO_3^--N consumed. Our Y_b value agrees with those reported from tests with similar NO_3^--N concentrations (Table 11). However, Moore & Schroeder (1971) showed that under steady-state flow-through conditions and NO_3^--N feed variation, Y_{NO3}-N decreases linearly with increasing NO_3^--N concentration to about 35 mg/l NO_3^--N, and remains constant thereafter. They attributed this relationship to a saturation effect, because some species of bacteria synthesize polysaccharide storage materials under nitrogen-limited conditions. Consequently, the process slows as more NO_3^--N is utilized, resulting in Y_{NO3-N} decreasing until NO_3^--N reaches the saturation level. Above 35 mg/l NO_3^--N, Y_{NO3-N} was found to be around 0.60 g VSS cells produced/g NO_3^--N consumed (Moore & Schroeder, 1971), which is similar to the value we found. Hence, we infer that tests in our study were conducted under saturation conditions. Indeed, batch tests started from a NO_3^--N concentration of 207 mg/l, and were interrupted close to the putative saturation limit.

Source	Y(Biomass VSS/ NO₃⁻-N)	Y(Biomass VSS/ Dissolved COD)	Y(Biomass COD/ Dissolved COD)
This study	0.69	0.29	0.41
Semon et al. (1997)	0.62	0.17 - 0.18	0.24 - 0.26
Jeris et al. (1977)	0.57	0.17	0.24
Moore & Schroeder (1971)	0.53 - 1.4	0.14 - 0.29	0.17 - 0.35
Coelhoso et al. (1992)	0.5 - 1.3	-	-
Stephenson & Murphy (1980)	1.0	-	-
Grady et al. (1999)	-	0.27	0.39

Table 11. Comparison of biological yield (Y_b) estimates from this study (batch reactor tests) and from the literature under NO_3^--N saturation conditions.

Y_{VSS} and Y_{COD} were determined to have values of 0.29 and 0.41 g/g, respectively. These values agree only with those reported by Grady et al. (1999), and are larger than values reported by other authors (Table 11). For example, Jeris & Owens (1975) suggested that between 15 and 20% of the methanol consumed is expected to be converted into cell mass, while Karnchanawong & Polprasert (1990) found this conversion to be between 20 and 28%. The difference could be explained by noting that Grady et al.'s (1999) values were obtained under conditions similar to those in this study, such as excess NO_3^--N, and continuously-stirred, batch tank reactors. Such conditions exploited the maximum potential for ATP formation under anoxic conditions, resulting in a higher yield for anoxic growth. In contrast, other authors (Table 10) derived their results from steady-state operating conditions in fluidized bed biological reactors, which have much higher denitrification rates than continuously-stirred tank reactors. Lower denitrification rates coincide with high solids production rates (Stephenson & Murphy, 1980), helping to explain the larger Y_{VSS} and Y_{COD} values observed in continuously-stirred tank reactor tests.

We also determined Y_{NO3-N} using denitrification reactor data in order to confirm the results and interpretations above, and also to characterize the behavior of the denitrification reactor under conditions of diurnal NO_3^--N variations. Our findings (Table 12) confirmed that the denitrification reactor worked at an NO_3^--N dose lower than saturation and resulted in a larger Y_{NO3}-N. Additionally, the largest yields were obtained for the afternoon measurements, i.e., the lowest NO_3^--N influent, regardless of experimental treatment or working stream flow.

The weighted average biomass production for the denitrification reactor (as VSS) was estimated at 20.2 kg VSS m³/day for the 6 Lpm working flow rate, and 15.8 kg VSS m³/day for 4 Lpm. The difference was probably due to the different percent of recirculation, which resulted in different working streams. As was suggested by the biofilter Y_{NO3}-N (i.e., approximately one), nitrogen removal values approximated those of VSS removal. Nitrogen removal was between 23.4 kg NO_3^--N m³/day for the 6 Lpm working flow, and 16.2 kg NO_3^--N m³/day for 4 Lpm. Our maximum nitrogen removal was higher than generally expected from denitrification for domestic wastewater treatment. For example, Coelhoso et al. (1992) obtained nitrogen removal of 5.4 to 10.4 kg NO_3^--N m³/day. Semon et al. (1997) suggested a maximum design loading of 6.4 kg NO_3^--N m³/day. Jeris & Owens (1975) reported nitrogen removal of 20.7 kg NO_3^--N m³/day in fluidized sand biological reactors. The higher removal rate in our study was probably because of the higher operating temperature, which drove

Treatments (Sampling time)	ΔVSS (mg/L)	ΔNO_x-N (mg/L)	$Y(\Delta VSS/ \Delta NO_3^-\text{-N})$ (g/g)	Parameter weighted average per treatment			
				VSS produced		NO_3^--N removal	
				kg/d	kg/m³ d	kg/d	kg/m³ d
1 & 2 (2:00 p.m.)	9.17	7.49	1.24				
1 & 2 (10:00 p.m.)	23.65	26.99	0.88	0.303	20.203	0.351	23.407
1 & 2 (6:00 a.m.)	27.50	29.40	0.94				
3 & 4 (2:00 p.m.)	6.50	4.18	1.58				
3 & 4 (10:00 p.m.)	18.84	19.57	0.96	0.237	15.797	0.243	16.198
3 & 4 (6:00 a.m.)	18.67	19.91	0.95				

Table 12. Biological yields (Y_{NO3-N}) data from denitrification reactor (steady state conditions) for each experimental working flow, VSS produced, and NO_3^--N removal.

higher reaction rates. Furthermore, these conditions could imply even higher nitrogen removal if recycling did not have to be employed in order to assure bed fluidization. Additionally, the reactor maintained a high VSS biomass concentration, around 38,000 mg/l. Under these conditions, the sand from the settled media on average represented only 17% of the settled bed volume in the reactor. This information would be critical for the design of sand denitrification filters to be operated under similar conditions.

3.10 Estimation of operations costs for wastewater treatment at BRA

Experimentation with the pilot station showed that by using these treatment strategies, water quality can be improved to the degree that treated effluent is safe for fish production. The commercial feasibility of effluent treatment and reuse, however, depends upon an assessment of benefits and costs. Estimation of the construction costs for a full-scale wastewater treatment station based on the pilot-scale design that we evaluated is beyond the scope of our study. Further, amortization of capital costs is highly variable among countries and times and hence is not well given to useful discussion. Operating costs, however, are estimable using data available to us. Operation of a scaled-up plant treating the entire BRA effluent of 2260 m³/day would require electricity, oxygen for ozone production, methanol as a substrate for dentrification, ferric chloride for flocculation, and labor for operations and maintenance.

We assumed an ozone dose of 0.1 g/l wastewater, about 15% higher than the dose applied in treatment 3, the most successful treatment during pilot station experiments. Results of the pilot-scale study showed that such an increase should be economically feasible. This dose represents a total of 226 kg O_3/day. At an average of 12.14 kWh consumed per kg of ozone produced, the energy required daily will be 2743.6 kWh. At a local market price of $0.04/kWh, the costs of producing ozone will be US$110/day. Considering that the electricity required to operate the station is 10% of the energy required to produce ozone, total costs for electricity would be $120.75/day.

With current technology, which is capable of transforming 12% of oxygen into ozone, 1652.87 m³ of oxygen will be needed to produce the daily amount of ozone applied (i.e., 1.17 m³ O_2/min). This volume is equivalent to 2257.6 kg O_2 (at 1.43 kg O_2/m³). Assuming oxygen recirculation and a supplemental loss of oxygen due to dissolution in water, oxygen consumption should be approximately 10% of the amount used. Hence, the cost of 225.8 kg of oxygen consumed daily will be $15.50/day, at a bulk price of $0.092/m³.

BRA releases 97 kg NO_3^--N with the effluent, at an average concentration of 43 mg/l. At an assumed methanol:NO_3^--N ratio of 3.2:1 required by denitrification, 310.4 kg methanol per day will be needed. At a methanol density of about 0.8 kg/l, 388 L of methanol will be needed. Industrial grade methanol has a 98% concentration, which results in a total requirement of 396 L. At a price of $0.50/L, the cost of methanol will be $198/day.

Jar tests showed that treatment with an average of 50 mg $FeCl_3$/l wastewater, followed by sand filtration, could reduce the TSS in the treated effluent well below 10 mg/l. At this dose, 113 kg $FeCl_3$/day would be needed. At an average market price of $300/metric ton $FeCl_3$, the daily cost should be $34.

The operation of a full-scale wastewater treatment station would require full-time labor, including weekends. This will add up to 56 hrs of labor per week, or 240 hrs per month. At an estimated wage of $10 per hour, labor costs will add $2,400 per month to costs.

Operation of the wastewater treatment system would minimize certain operations costs currently faced by BRA. Water temperature did not decrease by more than 1-1.5°C as water passed through the pilot station during the 3-4 hours of treatment. However, the pilot station was operated during the summer, and high environmental temperature could have affected the rate of temperature loss. Considering that the treatment station also would function during the winter and that water would be held in storage tanks before and after treatment, we assumed that temperature would decrease by as much as 7°C. Because BRA spends between $10,000 and $28,000 per month (depending on season) on fuel oil to heat water, 75% heat recovery would represent $14,250 per month in savings, from the current average of $19,000. That is, only $4,750/mo ($158.33/day) would be spent to bring replacement water to culture temperature, and $14,250 per month would represent savings.

Releasing the waste stream instead of treating and reusing it adds to BRA's operating costs as wastewater discharge bill from the city of $14,000/month ($467/day). The energy to pump replacement water from the wells adds $450/month ($15/day) to operations costs. The summation of these costs results in total estimated expenses of $595.83/day ($17,875/month) for consumable materials. Against this total, $33,450 is the actual average cost of heating the replacement water plus the municipal water treatment charge and the energy cost for pumping the water from the wells. Were the entire effluent reused, operations costs of wastewater treatment at BRA would be reduced by $15,575 per month. Hence, much of the economic determination of whether to go forward with wastewater treatment and reuse will depend upon construction costs and the amortization into ongoing payments.

Our findings regarding costs and savings for treatment of effluent were such that BRA has gone forward with investment in full-scale effluent treatment. While the treatability of aquaculture effluent and the costs structure for treatment are specific to a given operation, the approach we took for assessing treatability and for designing and evaluating a treatment train are general, and will have relevance to a range of recirculating aquaculture system operations.

4. Conclusion

Recirculating aquaculture systems have been developed to produce high-value species for year-round supply of markets, to free production from site-related constraints, to minimize environmental impacts of aquaculture, and to make efficient use of limited high-quality water supplies. Effective treatment and reuse of aquaculture effluent has been demonstrated at an experimental scale. We built and operated a pilot-scale wastewater treatment station at a large commercial recirculating aquaculture facility in order to evaluate treatment strategies for effluent recovery and reuse. The treatment train consisted of sedimentation, denitrification, ozonation, nitrification and chemical flocculation. We report the dynamics of alkalinity, pH, hardness and nitrogenous compounds through the treatment process. Alkalinity lost by water exchange was recovered due to nitrification, and alkalinity in treated effluent was 26-33% higher than initial alkalinity. pH increased from a mean of 7.21 in the initial influent to 7.60 in the final effluent after larger changes through the treatment train. Hardness decreased by approximately 10%, with the degree of decrease positively correlated with ozone dose and with associated removal of total suspended solids. Up to 96% of total Kjeldall nitrogen was removed, mostly as organics. Although ammonia was produced during ozonation, it was partially removed in the trickling filter, decreasing by 35-40% after treatment. Over 94% of NO_3^--N was removed by the treatment train, declining to 2.0-2.7 mg/l. The biological yield for denitrification, Y_b (g biomass volatile suspended solids/g NO_3^--N), was 0.69, and maximum nitrogen removal was 23.4 kg NO_3^--N /m^3-day. The nitrogen removal in the denitrification reactor was between 16 and 23 kg NO_3^--N /m^3-day. Nitrogen removal was higher than generally expected from wastewater treatment, in part because of the high temperature of operation, 28-30°C. We conclude that the pilot station design was effective for conserving alkalinity and hardness and for removing nutrients, and could be scaled up to treat and reuse the entire effluent stream. Should the system be scaled up, our results predict significant savings in operations costs, largely due to savings in energy required to heat the exchange water. While the treatability of aquaculture effluent and cost structure for treatment were specific to BRA, the approach we took to assess treatability and to evaluate the treatment train are general and will have applicability to a range of recirculating aquaculture system operations.

5. Acknowledgement

S.S. was supported by a Commercial Fish and Shellfish Technologies grant award to E.H. and by the Department of Fish and Wildlife Conservation at Virginia Polytechnic Institute and State University. We are grateful for access to the production facilities at Blue Ridge Aquaculture and to the water quality laboratory of the Department of Civil and Environmental Engineering at Virginia Tech University, and for the technical training and advice of Dr. Nancy Love and Julie Petruska.

6. References

APHA (American Public Health Association), American Water Works Association and Water Environment Federation (1998). *Standard Methods for the Examination of Water and Wastewater, 20th edition*. American Public Health Association, ISBN 0-87553-235-7, Washington, DC

Beltran, F.J.; Garcia-Araya, J.F. & Alvarez, P.M. (2001). pH sequential ozonation of domestic and wine-distillery wastewaters. *Water Research*, Vol. 35, No. 4, pp. 929–936, ISSN: 0043-1354

Buxton, G.V.; Greenstock, C.L., Helman, W.P. & Ross, A.B. (1988). Critical review of rate constants for reaction of hydrated electrons, hydrogen atoms and hydroxyl radicals (oOH/oO⁻) in aqueous solutions. *Journal of Physical Chemistry Reference Data*, Vol. 17, No. 2, pp. 513–886, ISSN 0047-2689

Coelhoso, I.; Boaventura, R. & Rodrigues, A. (1992). Biofilm reactors: An experimental and modeling study of wastewater denitrification in fluidized-bed reactors of activated carbon particles. *Biotechnology and Bioengineering*, Vol. 40, No. 5, pp. 625 – 633, ISSN 0006-3592

Grady, C.P.L.; Daigger, G.T. & Lim H.C. (1999). *Biological Wastewater Treatment, 2nd edition*. Marcel Dekker, Inc., New York, ISBN 0824789199

Grasso, D. & Weber, W.J. (1988). Ozone-induced particle destabilization. *Journal of the Water Works Association*, Vol. 80, No. 8, pp. 73–81, ISSN 0003-150X

Jeris, J.S. & Owens, R.W. (1975). Pilot-scale, high-rate biological denitrification. *Journal of the Water Pollution Control Federation*, Vol. 47, No. 8, pp. 2043–2057, ISSN 0431303

Jeris, J.S.; Owens, R.W. & Hickey, R., (1977). Biological fluidized-bed treatment for BOD and nitrogen removal. *Journal of the Water Pollution Control Federation* Vol. 49, pp. 816–831, ISSN 0043-1303

Karnchanawong, S. & Polprasert, C. (1990). Organic carbon and nitrogen removal in attached-growth circulating reactor (AGCR). *Water Science and Technology*, Vol. 22, No. 3/4, pp. 179–186, ISSN 0273-1223

Kirk, B.S., McNabney, R. & Wynn, C.S. (1975). Pilot plant studies of tertiary wastewater treatment with ozone. *Ozone in water and wastewater treatment*. In: F.L., Evans, editor. Pp. 61–82. Ann Arbor Science Publishers, Inc., ISBN 0250975238, Ann Arbor, MI

Lin, S.H. & Wu, C.L. (1996). Removal of nitrogenous compounds from aqueous solution by ozonation and ion exchange. *Water Research*, Vol. 30, No. 8, pp. 1851–1857, ISSN: 0043-1354

Metcalf and Eddy, Inc. (1991). *Wastewater Engineering: Treatment, Disposal and Reuse, 3rd edition*, McGraw-Hill, ISBN 0070416907, New York

Millamena, O.M. (1992). Ozone treatment of slaughterhouse and laboratory wastewaters. *Aquacultural Engineering*, Vol. 11, No. 1, pp. 23–31, ISSN 0144-8609

Moore, F. & Schroeder, E.D. (1971). The effect of nitrification feed rate on denitrification. *Water Research*, Vol. 5, pp. 445–452, ISSN 0043-1354

Parker, D.S. & Richards, T. (1986). Nitrification in trickling filters. *Journal of the Water Pollution Control Federation*, Vol. 58, No. 9, pp. 896-902, ISSN 0043-1303

Razumovskii, S.D., & Zaikov, G.E., (1984). *Ozone and its reactions with organic compounds*. Elsevier Science Publishers B.V., ISBN 0444423699, Amsterdam

Rosenthal, H. & Otte, G. (1979). Ozonation in an intensive fish culture recycling system. *Ozone: Science and Engineering*, Vol. 1, pp. 319 – 327, ISSN 0191-9512

Rueter, J. & Johnson, R. (1995). The use of ozone to improve solids removal during disinfection. *Aquacultural Engineering*, Vol. 14, No. 2, pp. 123–141, ISSN 0144-8609

Sandu, S. (2004). Evaluation of ozone treatment, pilot-scale wastewater treatment plant, and nitrogen budget for Blue Ridge Aquaculture. Ph.D. dissertation. Virginia Polytechnic Institute and State University, Blacksburg, Virginia, USA

Sandu, S.; Brazil, B. & Hallerman, E. (2008). Efficacy of a pilot-scale wastewater treatment plant upon a commercial aquaculture effluent: I. Solids and carbonaceous compounds. *Aquacultural Engineering*, Vol. 39, No. 1, pp. 78–90, ISSN 0144-8609

Semon, J.; Sadick, T., Palumbo, D., Santoro, M. & Keenan, P. (1997). Biological upflow fluidized bed denitrification reactor demonstration project – Stamford, CT, USA. *Water Science and Technology*, Vol. 36, No. 1, pp. 139–146, ISSN 0273-1223

Stephenson, J.P. & Murphy, K.L. (1980). Kinetics of biological fluidized bed wastewater denitrification. *Progressive Water Technology*, Vol. 12, pp. 159 – 171

Summerfelt, S.T. (2003). Ozonation and UV irradiation – an introduction and examples of current applications. *Aquacultural Engineering*, Vol. 28, No. 1, pp. 21-36, ISSN 0144-8609

van Rijn, J. & Rivera, G. (1990). Aerobic and anaerobic biofiltration in aquaculture. Unit-nitrite accumulation as a result of nitrification. *Aquacultural Engineering*, Vol. 9, No. 2, pp. 217–234, ISSN 0144-8609

Wang, G.S. & Pai, S.Y. (2001). Ozonation of dissolved organic matter in biologically treated wastewater effluents. *Ozone: Science and Engineering*, Vol. 23, pp. 351–358, ISSN 0191-9512

Part 3

Energy, Metals and Instructive Case Study

9

Metals in Crawfish

Joseph Sneddon and Joel C. Richert
Department of Chemistry, McNeese State University,
Lake Charles, Louisiana
USA

1. Introduction

1.1 Overview

Much of the information in the first nine sections of this chapter was derived and condensed from the Louisiana Crawfish Production Manual (McClain et al, 2007). The harvesting of crawfish (or crayfish) for human consumption has become an important industry in several areas throughout the world. In the United States, crawfish are common in Louisiana (located in the southern United States on the Gulf of Mexico) and throughout the southeastern states. Crawfish (McClain, 2005) are well suited for habitats that have seasonal flooding and drying, especially when the dry periods occur in the summer and fall. Periods of flooding allow the crawfish to feed, grow and mature. Dry periods help aeration of sediments, reduce the abundance of predators and allow for establishment of vegetation, which serves as cover and food resources when flooded. Crawfish are common in parts of northeastern Mexico, can be found extensively in southeastern Asia and are grown commercially in China. In the 1970s, crawfish were introduced into Portugal and Spain and thrived in the rivers and estuaries to the extent that commercial harvesting now contributes significantly to the food supply in these regions (Alcorlo et al, 2006) (Maranhao et al, 1999). In Louisiana there are numerous varieties of crawfish but the two dominant species are the red swamp crawfish (*Procamburas clarkii*) and the white river crawfish (*Procamburas zonangulas*). Both species dominate harvests in natural habitats and on crawfish farms, therefore, unless otherwise noted these two species are the species of crawfish to which the authors refer.

1.2 History

In Louisiana, crawfish has been a part of the human diet for centuries but the commercial sales of crawfish began only in the late 1800s. As the demand for crawfish outgrew the harvest from the wild, the crawfish aquaculture industry was born. Early crawfish aquaculture consisted only of re-flooding rice fields after the harvest to produce food for the farm family. Excess production from these fields was sold to nearby consumers. Over time the demand and supply of crawfish continued to grow and by the 1960s commercial processing of crawfish began. The peeling and packaging of crawfish tail meat allowed for the establishment of regional markets. Commercial processing required a steady supply of crawfish to meet demand. This increased demand was greater than could be supplied by harvests in natural habitats and lead to a great increase crawfish aquaculture in Louisiana. In the United States there are few commercial producers of crawfish outside of Louisiana.

Louisiana produces about 90 percent of the nation's crawfish, 70 percent of which is consumed within the state (McClain, 2005).

2. Crawfish

2.1 Anatomy

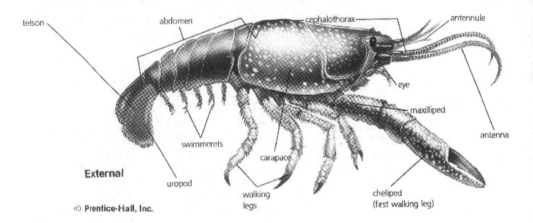

www.thomson.fosterscience.com

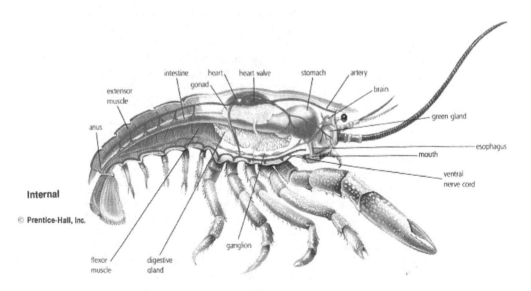

www.thomson.fosterscience.com

2.2 Biology

Both red swamp and white river crawfish are considered a temperate species of crustaceans that are able to tolerate moderately cold winters. There are some differences of note in these two species. White river crawfish spawn in the fall of the year and produce larger but less numerous eggs than do the red river crawfish which spawn year around. Red river crawfish have higher feeding rates at temperatures above 30° C. Additionally, the red swamp crawfish are more tolerant to the low dissolved oxygen content that sometimes occurs in crawfish ponds. This may be the reason that the red swamp crawfish dominate most crawfish production ponds. The two species are very similar in appearance but can be differentiated between by an experienced person. Red swamp crawfish is the preferred species because of this species greater reproductive potential and a more prolonged reproductive season. There is no evidence of natural hybrids between the two species.

2.3 Life cycle

Both species live about 2 years, have high juvenile survival rates and have reproductively active and inactive periods. After mating in open water, the female crawfish then stores the sperm in an internal receptacle. Egg development in the females takes place internally. Developing eggs in the ovary increase in size and change from a light color to black. More than 500 mature eggs are expelled and attached to the swimmerets under the tail of the crawfish. Fertilization of the eggs takes place externally when the female releases the stored sperm. During egg production the female builds a burrow and spawning takes place in the burrow. The incubation period for the eggs is approximately three weeks. The hatchlings remain attached to the female's swimmerets through the first two molting phases. After detaching from the mother, the young crawfish resemble adult crawfish and begin to feed. The female and young crawfish remain in the burrow for several weeks but then must leave because little food is available in the burrows. Pond flooding or heavy rainfall will encourage the crawfish to emerge sooner rather than later. Reproduction is somewhat synchronized in pond-raised crawfish because rice fields are flooded in the fall after the rice is harvested to coincide with the optimum time for reproduction. While red river crawfish spawn at anytime, autumn is the best time for white river crawfish to spawn. Peak production can best be attained when reproduction is properly timed in autumn.

Crawfish, like all crustaceans, must molt periodically to increase in size. Growth rate is dependent on variables such as: water temperature, oxygen levels, population density and food quality and quantity. Optimum harvest size can be reached in as little as 7 to 9 weeks, but usually takes 3 to 5 months after hatching. Crawfish molt approximately 11 times before reaching maturity. Though there are five major stages in the molt cycle, it is a continuous process. The inter-molt stage is the phase in which the exoskeleton is hard and fully formed. During this stage the crawfish feed to increase their energy and tissue reserves. In the pre-molt stage an underlying soft exoskeleton forms and begins re-absorbing calcium for the older exterior. During molting feeding stops and temporary shelter is sought. Molting takes place in a matter of minutes as the old exoskeleton splits between the carapace and the abdomen on the dorsal surface. The crawfish then withdraws by flipping its tail several times. After emerging from the old exoskeleton the crawfish is in the "soft" phase during which the new exoskeleton expands to a larger size than the previous exoskeleton. Calcification of the new exoskeleton takes place in two phases. During the first phase the calcium stored by the crawfish in two hard gastroliths on each side of the stomach is transported to the new shell. The second phase of the hardening occurs as calcium is absorbed from the water. Once crawfish molt to a reproductively active stage, growth ceases. A medium size crawfish has a mass of about 25 grams and a head to tail length of 10 centimeters. Mature crawfish can be identified by dark coloration, enlarged claws and harden sexual structure. Adult males develop prominent hooks on the base of the third and fourth pair of walking legs. Females will mate several times after molting to the mature stage.

2.4 Crawfish borrows

Crawfish dig unbranched vertical burrows usually less than one meter deep. The burrows are used for purposes other than reproduction. The burrows serve as refuge from predators and provide a moist environment necessary to survive dry periods. Crawfish burrows are built over several days time by an individual crawfish and the burrow diameter is dependent on the size of the crawfish. At the bottom of the burrow is chamber slightly larger than the crawfish. The water level in a burrow is dependent on the moisture conditions in the surrounding soil. The entrance to a completed burrow is sealed with a mud plug. Crawfish burrows often have a soil stack or chimney above ground which is formed by the excavated dirt. Once sealed in the burrow the crawfish is confined to the burrow until the plug is softened by heavy rainfall or pond flooding.

2.5 Population dynamics

In most forms of aquaculture, known numbers and sizes of juveniles are used to stock the ponds. Stocking with juveniles is not used in Louisiana crawfish aquaculture. Population within a pond depends upon reproduction from either stocked or already present mature crawfish. Density of the population depends upon brood stock survival, successful reproduction and survival of the offspring. Density can be adversely affected by factors over which the farmer has no control such as: by low oxygen levels, predators or pesticide exposure from nearby farming operations. Research has revealed that density greater than 15 crawfish per square yard results in slow growth rates and smaller sized crawfish at maturity. These factors have made control over population levels one of the most elusive aspects of crawfish aquaculture.

2.6 Nutrition

Crawfish are omnivores, detritivores (consumers of decomposing organic matter) and most recently as obligate carnivores because it has been found that they "require" some animal matter in their diet to optimize growth. A crawfish diet includes living and decomposing plant matter, seeds, algae, microorganisms, epiphytic organisms, and many invertebrates such as worms, snails and insects. They will also eat small fish and other smaller crawfish. Living and decomposing plants are often the most abundant food source in a crawfish pond, yet contribute very little directly to the nourishment of crawfish. These plants do supply most of the nutrients in the ecosystem of a crawfish pond. The decomposing plants and its associated microorganisms (commonly referred to as detritus) are consumed because it has high food value than living plants. The amount of detritus that can be utilized by crawfish as the mainstay of their nutrition is limited. This microbe rich detritus is the main food source for insects, worms, snails and some small vertebrates that then consumed by the crawfish. These organisms provide high quality nutrition for crawfish.

Supplemental feeds are not often utilized in Louisiana crawfish aquaculture ponds. The primary means of providing nutrition to crawfish in aquaculture is by establishing and managing a forage crop in the crawfish ponds. Once the ponds are flooded in the fall the forge crop and detritus produced by it must provide a constant and continuous supply of nutrients for the food web on which the crawfish relies for its nutrition.

3. Crawfish aquaculture ponds

3.1 Location and layout

Ponds are located in open flat areas in soils with a sufficient amount of clay in the soil. Soils that can be shaped into a ball have enough clay to be suitable. The clay content is important to hold water during flooding and to maintain the integrity of the crawfish burrows. Water resources for periodic flooding of the ponds are also necessary. The pond must be at an elevation above the water levels of surrounding canals and ditches. On farms where crawfishing is rotated with rice crops, several factors need to be considered. Ponds must have adequate all-weather access because aquaculture of crawfish requires almost daily harvesting and pond management during the January through June harvesting period. Because crawfish farming is labor intensive, only 10 to 50 percent of a farmer's rice acreage will be selected for crawfish aquaculture. Farmers usually do not rotate the best rice producing fields with crawfish farming.

There is no standard size for a crawfish pond. A size range of 10 to 40 acres is prevalent in Louisiana. Most crawfish producers manage fewer than 150 total acres. For the purpose of trapping, long fields with few levee crossing are the most efficient. Consideration must be given to other nearby farming operations where the aerial application of pesticides may contaminate downwind crawfish ponds.

3.2 Design and construction

Whether constructing a pond for permanent crawfish production or using an existing rice field for rotation with crawfish, many considerations need to be taken into account. The perimeter levees should be 3 meters wide at the base to prevent leakage from crawfish burrowing. Levees that are 1 meter tall are adequate to maintain the minimum of 20 to 25 centimeters of water necessary for crawfish production. The fall of the land should not exceed 15 centimeters between levees. Fields with steeper grades result in water depth variations that hamper forage growth and reduce harvesting efficiency. Drains for the field should be sized based on field size, projected rainfall and irrigation pumping capacity. One 25 centimeter diameter drain is sufficient per 10 acres of pond. Interior levees within the pond should be 2 meters at the base and be 15 centimeters above the water level in the pond. These interior levees should be spaced 50 to 100 meters apart to facilitate water circulation. Wide or deep interior ditches should be avoided if possible. These ditches provide the least resistant flow of water within the pond which can lead to poor circulation in the rest of the pond. Poor circulation of water can lead to stagnation of the water and cause lower production of crawfish. Deep areas within the field will also make periodic draining the fields difficult.

Crawfish pond with a row of traps/ Courtesy of Joel Richert

4. Production systems

4.1 Monocropping system

Monoculture or "single crop" aquaculture of crawfish is used on small farms or in areas where agriculturally marginal land is available. Ponds devoted to numerous production cycles are typically used for this strategy. The size of these operations range from 300 acres with little management or input to small 15 acres sites that are highly managed. The big advantage to monocropping is that farmers manage the acreage for maximum crawfish production without concerns of other crops, such as pesticide contamination and seasonal limitations. The yields from this kind of system range from 100 kg/ac in large low input ponds to more than 600 kg/ac in small intensively managed ponds. "Permanent" ponds yields will generally increase over a period of three to four consecutive years of production.

Rice is the standard forage crop used in a monocropping system. It is planted in the summer months after the crawfish harvesting season. Emphasis is on the growth of stem and leaf production for use as forage. The rice planted is not harvested for grain at maturity, rather it is allowed to die and begin decomposition. One of the advantages of permanent ponds is earlier, intense harvesting of crawfish. The early harvesting of crawfish is economically important because early harvest is almost always rewarded by high market prices. The disadvantages to monocropping include: 1) the need to construct dedicated ponds, costs must be amortized over only one crop, and crawfish overcrowding can occur after a few years which leads to stunted, low-priced crawfish that are difficult to market. Production schedules for monocropping can vary, but generally follow the schedule on Table 4.1.

4.2 Crop rotation systems

In rotation systems, crawfish is alternated with a plant crop in order to grow two crops in one year. Rotational strategies have the advantage of efficient use of labor, farm equipment

	Summary of Major Crawfish Production Strategies With Common Practices by Month.		
		Crop Rotational Systems	
Months	Crawfish Monoculture	Rice-Crawfish-Rice	Rice-Crawfish-Fallow or (Rice-Crawfish-Soybean)
Jul - Aug	Forage crop planted or natural vegetation allowed to grow	Rice crop harvested in August and stubble managed for regrowth	Rice crop harvested and stubble managed for regrowth
Sep - Oct	Pond flooded and water quality monitored and managed	Pond flooded in October and water quality monitored and managed	Pond flooded in October and water quality monitored and managed
Nov - Dec	Harvest when catch can be economically justified	Harvest when catch can be economically justified	Water quality monitored and managed
Jan - Feb	Crawfish harvested 2-4 days per week according to catch and markets	Crawfish harvested 2-4 days per week according to catch and markets	Crawfish harvested 2-4 days per week according to catch and markets
Mar - Apr	Crawfish harvested 3-5 days per week according to catch and markets	Crawfish harvested 3-5 days per week until late April, then pond drained and readied for planting	Crawfish harvested 3-5 days per week according to catch and markets
May - Jun	Crawfish harvested until catch is no longer justified; then pond drained	Rice planted in May and rice crop managed for grain production	Pond drained and soybeans planted or harvest proceeds as long as catch is feasible; pond then drained and left fallow
July - ...	Repeat cycle	Repeat cycle	Harvest soybeans in October, plant rice in March/April, stock crawfish in May, repeat cycle

Louisiana Crawfish Production Manual

Table 4.1

and land. Additionally, some fixed costs and the costs of a plant crop can be amortized over two crops rather than just one. Rotational systems can vary based on the type of land, crops already in production on a farm or past experience of the farmer.

4.3 Rice-crawfish-rice rotation

Rotation of rice and crawfish takes advantage of the seasonality of each crop which allows for the production and harvest of each crop in one year. Rice is grown and harvested in the summer months while crawfish grown during the fall, winter and spring in the same field in a twelve month period. In rice rotation systems the initial stocking takes place 4 to 7 weeks after the rice is planted. After the rice is harvested in July or August, the residual rice stubble is fertilized with a nitrogen based fertilizer in order to establish a re-growth crop of rice for forage. After a fall flooding of the pond, management practices are similar to the monocropping system. (See Table 4.1) The major disadvantage to this rotational system is that neither crop can be managed to maximum efficiency. In Louisiana rice yields are maximized when it is planted in early spring. Draining the ponds and planting rice in early spring shortens the crawfish harvest season. Delaying the planting of rice will adversely affect rice yields. Care must be taken in the use of pesticides and herbicides that are helpful to the rice crop but may be harmful to the crawfish crop. Yields for each crop can vary greatly depending upon the management emphasis. Generally the crop that is given preferred treatment has higher yields at the expense of lower yields for the less preferred crop.

4.4 Crawfish-rice-fallow (or rice-crawfish-other plant crop)

The big difference in these rotational strategies is that rice is typically not cultivated in the same field in consecutive years in order to control rice diseases and weeds. Like rice-crawfish-rice rotation crawfish culture follows rice cultivation; therefore crawfish production does not take place in the same ponds from one year to the next. In the first mentioned rotation method, the field will be left fallow for a year following the crawfish harvest. Under the second method of rotation, if soybeans or another crop is planted after the crawfish harvest, then three crops per field can be harvested in a two year time frame. These two approaches require sufficient acreage to allow staggered crops in different fields within the farm. This rotational method is used on much of the acreage used in crawfish aquaculture in Louisiana because it has advantages over rotation within the same field. Each of the crops in this type of rotation can be better managed because each crop has the necessary time to maximize growth and harvest. Specifically for crawfish production, the crawfish ponds do not have to be drained in late spring to rice. This yields a lengthening of the crawfish harvest season into early summer because soybeans or other crops can be planted later in the year. The disadvantages of this rotation system include: the need to re-stock crawfish each year, lower population densities and the late season harvest frequently is frequently plagued by low market prices for the crawfish.

5. Stocking

5.1 General guidelines

Crawfish farming relies on reproduction of the resident adults to sufficiently populate the ponds for each harvest. Established ponds seldom have a need to be re-stocked because the crawfish that remain after the harvest serve as brood stock for the next year. Stocking is usually only necessary in: 1) new ponds, 2) after a fallow year or crop rotation other than rice, 3) after severe drought or 4) after extensive levee renovation.

5.2 Stocking considerations and procedures

Red swamp crawfish is the preferred species for stocking because of their longer reproductive season. The size of the mature broodstock is of little concern, because even though large crawfish produce a higher number of offspring, there would be fewer of them per purchased kilogram of crawfish. The time for stocking depends upon the reason for stocking. April to July is the best time to stock a new pond in a monocropping system or restock a pond at has skipped a season. When re-stocking a rice field, stocking should be done about 45 days after planting when the rice plants are large enough to withstand the crawfish without damage and when the need for harmful pesticide applications have passed. Careful handling of the broodstock from trapping to release is important. Keeping the crawfish clean, moist, and at temperatures around 20° C. This is best accomplished by avoiding direct sunlight, wind, and by covering the sacks of crawfish with wet tarpaulins or burlap sacks. When stocking, crawfish should be dumped directly in the water. Because crawfish are mobile, there is no need to equally spread them over the entire pond. The crawfish should however be stocked in each section of a large pond. The temperature of the pond being stocked should not differ greatly from the temperature of the crawfish themselves at the time of stocking. Stocking rates of 25 to 30 kg/acre is generally recommended but many farmers stock up to 40 kg/acre if post-stocking survival is unpredictable. The female population should exceed 50 percent because a male crawfish can mate with more than one female. Survival rate of the broodstock is extremely important to the stocking process. Water temperature, dissolved oxygen levels and water levels in the stocked pond must be maintained.

6. Forage management

6.1 Forages

Establishment of a forage-based production system is important in order to produce the complex ecological community necessary to provide a high quality food source for the crawfish. The food sources rely on a continuous influx of plant matter that is in turn consumed by bacteria and other microorganisms. The detritus produced by these decomposers is the fuel for the food web on which the crawfish rely. The forage crop must be able to produce adequate portions of plant material on a consistent basis throughout the growth and harvest seasons. Overproduction of plant material at one time is wasted because it cannot be stockpiled for later use. Decomposition of large amounts of material in a short span of time can also lead to oxygen depletion in the water. Too little plant material can lead to insufficient detritus to support the food web. Selected agronomic crops are the most effective forage resources for crawfish. These crops are most effective because of their ability to flourish in the flooded pond environment and their predictable plant material yield.

6.2 Types of forage

Rice is by far the most often used forage plant in Louisiana. Rice is semi-aquatic and grows well in flooded crawfish ponds. Many varieties of rice have been used as forage. In 2004, Louisiana State University released the first rice specifically developed for use in crawfish monocropping. "Ecrevisse" rice exhibits much greater forge biomass production, greater growth under the extended flood conditions of a crawfish pond and has an improved ability for post-winter re-growth than the commonly used domestic rice varieties. A sorgam-sudangrass hybrid that is commonly used by cattlemen for grassing and hay, is a well suited alternative forage crop for crawfish monocropping systems.

7. Water quality and management

7.1 Overview

Water quality is influenced by both environmental and biological factors. Some environmental factors such as rainfall and temperature are beyond control. Factors such as what type of vegetation planted, when planting occurs and how the vegetation is managed are within the controls of the producer and can affect water quality. Additionally, maintaining optimum water levels and the timing of flooding ponds can have a positive effect on water quality.

7.2 Quantity and supply

Surface and subsurface water are both used for flooding in crawfish aquaculture. Surface water that is pollution-free and free of predatory fish is cheaper than subsurface water but is usually not reliable as to quality or quantity. Wells provide predator-free water on demand but require a large investment and reoccurring pumping costs. Pumps are usually powered by diesel engines. Subsurface water has no oxygen and must be aerated before entering the pond. Subsurface water must also be monitored for high iron content and hydrogen sulfide. A pumping capacity of 300 to 400 liters per minute per surface acre is ideal. This pumping rate is sufficient to exchange all the water in a pond in 4 to 5 days. Flushing a pond completely is important in the early fall when water is flooded on to the vegetation. During the spring, warm weather causes rapid plant decay which causes a high demand for dissolved oxygen in the water. Low oxygen levels in the water cause high mortality rates and stress which reduces growth. The ability of the farmer to flush the ponds of oxygen deficient water in a timely manner is important. Few crawfish farmers have the capacity to supply water in the ideal quantities, therefore following an intense management plan can make up for lack of pumping capacity. Filling ponds to one-half normal depth will enable a pump with a smaller pumping capacity to flush the pumps quickly enough when necessary. However, if less than optimal water depth is utilized, more intensive monitoring of water quality in necessary because smaller volumes of water can change characteristics quicker than larger volumes of water.

7.3 Quality of water

Water quality variables include: 1) total hardness, 2) total alkalinity, 3) pH, salinity and other dissolved materials, and 4) dissolved oxygen. Dissolved oxygen is by far the most important because low oxygen is responsible for more crawfish mortality than any other factor. Temperature of the water is a factor because warm water cannot hold as mush oxygen as cooler water. In water that increases from 21° to 26° C the rate of oxygen use due to decomposition doubles. The source water after aeration should have a pH in the 6.5 to 8.5 range and total hardness and alkalinity should range from 50ppm to 250 ppm in calcium carbonate. Most water and soil in Louisiana that is used for crawfish production meets these parameters. Crawfish are fairly tolerant to salt water, however areas along the gulf coast that are subject to salt water intrusion should not be used as crawfish farms. Young hatchlings will die at 15 ppt salinity and juveniles die at 30 ppt salinity after a week of exposure.

8. Harvesting

8.1 Overview

The harvesting of crawfish utilizes baited traps that are periodically emptied. Trapping may begin as early as November and continues through the harvest season which usually ends by late June. In Louisiana, two-thirds of the crop is harvested from March through early June. In 2008, the most recent year in which accurate numbers are available, approximately 58 million kg of crawfish were harvested from aquaculture operations in Louisiana. The value of 2008 crawfish farm harvest was 115 million US dollars (Isaacs and Lavergne, 2010). Trapping is responsible for over half of the production expenses. The cost of bait and labor are the major harvesting costs. Crawfish yield within a pond can vary greatly from day-to-day and is governed by many factors. Water temperature, crawfish density, mass molting and weather are some of the variables that cause fluctuations in harvest.

8.2 Traps and baits

The "pyramid trap" is currently the industry standard in Louisiana. (Figure 8.1) Traps are constructed of 3/4-inch or 7/8-inch mesh wire formed into a three-sided pyramid. A 15 cm plastic collar on the opening at the top acts as handle. Traps are 60 cm wide and from 1.1 m to 1.4 m tall with one-way openings in the three bottom corners of the trap. Crawfish can easy enter the baited trap but cannot exit through the openings. The traps are secured in place on the bottom of the pond by rod that runs through the trap into the mud. The bait used in the trap is often the single highest expense in crawfish production. Bait cost depends upon the type, amount used, number of traps and trapping frequency. Various types of small fish, butchered larger fish or formulated baits are the most common baits used. Early in the season the fish baits are most commonly used because the strong scent of the fish is better at attracting crawfish in the cooler waters of that time of year. As the water warms, less expensive formulated baits are used. The formulated baits are densely packed cylinders (approximately 3 cm in diameter and 7 cm long) of cereal grains, grain by-products, favoring and a binder.

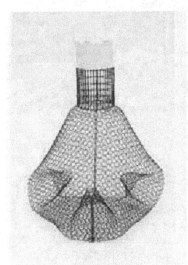 

Louisiana Crawfish Production Manual / Courtesy of Joel Richert

8.3 Trapping machinery and strategies

The most common method of collecting the harvest from the traps is by use of a specially designed paddle boat driven by a gasoline engine. The aluminum boats have flat bottoms and are typically 4.5 to 6 meters long and 1.5 to 2 meters wide. The 12 to 24 horsepower air-cooled engine drives a hydraulic pump to propel a metals cleated wheel attached to the boat. The boat is equipped with a sacking table on which the contents of each trap are dumped. Two or more sacks (capacity approximately 18 kg. each) temporarily attached to the table hold the crawfish as they are harvested. Many tables are designed to cull smaller crawfish and to remove bait and other debris from the catch. After dumping a trap, the operator re-baits the trap and returns it to the pond. The boat is in continuous motion and does not stop unless a problem occurs. A single operator can service about 150 to 200 traps per hour. Traps are placed in rows with 15 to 20 meters between rows and individual traps. The traps are emptied and baited usually three to four days per week. If prices and demand are high, traps may be run more often.

Courtesy of Joel Richert

9. Marketing

9.1 Live crawfish

Crawfish are sold live by farmers to wholesalers and live crawfish comprise most of the retail sales to consumers. The most common method of preparing crawfish for consumption, whether in households or in restaurants, is by boiling in well seasoned water. In times when supply exceeds demand, producers of large crawfish receive a better price for their crop. In times of over-supply, large crawfish are sold for immediate consumption and smaller crawfish are processed for later consumption.

9.2 Processed crawfish

A portion of the annual crop is processed and sold as fresh or frozen abdominal (tail) meat. Processing involves cooking, peeling and deveining the tail meat. The amount of tail meat obtained from a crawfish is approximately 15 percent of the live weight of the whole crawfish. The prepared crawfish meat is then sold in 1 pound (.45 kg) plastic bags. The processed crawfish meat is used to make jambalaya (a seasoned rice dish), etouffe (a stew-like preparation served over rice), or any other number of "cajun" dishes for which Louisiana is famous.

10. Environmental

10.1 Crawfish use as a bio-indicator

This section, as well as, sections 10.2 and 10.3 was taken from a literature review written by the authors of this chapter entitled *Determination of Inorganics and Organics in Crawfish*

(Richert and Sneddon, 2008). Numerous environmental studies have been done using crawfish as a bio-indicator to monitor pollution. Crawfish have been useful as vectors to monitor contaminants in water and soil because they are prolific, relatively sedentary, easily recognizable and have a reasonably long life span. Additionally, they are in constant physical contact with the water and surrounding soil. Their position in the food web is high enough that some biomagnification can occur from eating contaminated organisms from lower positions in the food web. Three recent studies have shown that crawfish can be useful in monitoring the levels of a variety of metals. Schmitt looked at the potential ecological and human risks associated with metals in fish and crawfish from activities in the Tri-States Mining District (TSMD) in Northeast Oklahoma (Schmitt et al, 2006). Crawfish and six species of frequently consumed fish were collected in 2001-2002 from the Spring River and Neosho River which drain the TSMD. Whole crawfish were analyzed in composite samples. Metals concentrations were found to be higher in the samples from the sites most heavily affected by the mining and were lower in the reference samples tested. The levels of Pb, Cd, and Zn exceeded current acceptable levels. Human consumption of crawfish from the area was restricted based on the results of this and previous studies. Monitoring contamination caused by a release of pollutants in 1998 in the Aznalcollar-Los Frailes, Spain mining region was the subject of a study (Sanchez Lopez et al, 2004). Immediately following the spill there was much destruction to the aquatic ecosystem of the Guadiamar River. Inductively coupled plasma-mass spectroscopy (ICP-MS) was used to measure levels of Cu, Zn, Pb and Cd. American red crawfish *(Procamburas clarkii)* were collected from various sites within the affected area. The crawfish collected in the areas near the toxic release showed much higher levels of the tested metals than did the crawfish collected from areas not directly affected by the spill. Additionally, a translocation experiment using red swamp crawfish *(Procamburas clarkii)* at different sites along the Guadiamar River was carried out in order to determine the ability of this species as a bio-indicator of heavy metal (Cd, Cu, Zn, Pb, and As) contamination (Alcorlo et al, 2006). Caged uncontaminated crawfish were placed at three different sites which had different levels of contamination. The crawfish were then harvested after six days and again after twelve days. Analysis by ICP-MS showed that in as little as six days the crawfish were already accumulating the metals. The study also showed that the metals not involved in crawfish metabolism tended to increase with higher exposure and longer exposures.

10.2 Inorganics or metals

Inorganics or metals are most frequently determined in crawfish using spectrochemical techniques such as flame and graphite furnace atomic absorption spectrometry (FASS, GFAAS), inductively coupled plasma-optical emission spectrometry (ICP-OES), and most recently inductively coupled plasma mass spectrometry (ICP-MS). It is beyond the scope of this chapter to describe these widely used and accepted techniques and the reader is referred to several texts that describe principles, instrumentation, and use of these techniques. However, in most cases these spectrochemical techniques perform best when the sample is in solution form, preferably in an aqueous or slightly acidic form. Typically microwave digestion is using nitric acid is used to put the analyte in solution. The toxic metals such as cadmium, mercury, and lead have been the most widely determined in crawfish.

10.3 Reviewed studies

Following a gold mine disaster in Romania, heavy metal content (Pb, Cd, Cr, Ni, Hg, As) was determined in silver carp, crawfish, sediment, and water (Francuski et al, 2000). Samples were collected from the Tisa River on February 28, 2000, upstream from the dam near Novi Becej in Rumania. Results showed heavy metals in all samples were significantly increased especially the Cd concentration in crawfish, which was three times higher than the maximal allowed concentration.

A study was performed to investigate direct and direct plus trophic contamination routes of crawfish *(Astacus astacus)* by inorganic Hg(II) or methylmercury (MeHg) (Simon and Boudou, 2001). Direct exposure was based on low contamination conditions, 300 and 30 ng/L in the dissolved phase, respectively, during 30 days at 208C. Trophic exposure was based on daily consumption of the Asiatic clam *(Corbicula fluminea)*, previously contaminated during 40 days with similar exposure conditions. The Hg concentrations in the bivalves were very similar: 1,451+287 ng/g for Hg(II) and 1,346+143 ng/g for MeHg. In the crustaceans, Hg bioaccumulation was determined at the whole-organism level and in eight organs (gills, stomach, intestine, hepatopancreas, tail muscle, green gland, carapace, and hemolymph), after 15 and 30 days of exposure. Analysis of the results showed marked differences between Hg(II) and MeHg accumulation in favor of MeHg: for the direct route, the ratio between metal concentrations was close to 8; for the trophic route, no significant increase in Hg accumulation

was observed for Hg(II) even when the ratio between Hg concentration in the direct plus trophic contamination route and Hg concentration in the direct contamination route was 1.6 for MeHg, with an estimated trophic transfer rate close to 20%. Mercury organotropism was also specifically connected to the exposure conditions, especially at the biological barrier level according to the route of exposure: gills and carapace for the direct route and digestive tract including hepatopancreas for the trophic route.

Another study evaluated potential human and ecological risks associated with metals in fish and crawfish from mining in the Tri-States Mining District (TSMD) in northeast Oklahoma (Schmitt et al, 2006). Crawfish *(Orconectes spp.)* and fish of six frequently consumed species (common carp *(Cyprinus carpio)*; channel catfish *(Ictalurus punctatus)*; flathead catfish *(Pylodictis olivaris)*; largemouth bass *(Micropterus salmoides)*; spotted bass *(M. punctulatus)*; and white crappie *(Pomoxis annularis)* were collected in 2001–2002 from the Oklahoma waters of the Spring River (SR) and Neosho River (NR), which drain the TSMD. Samples from a mining-contaminated site in eastern Missouri and from reference sites were also analyzed. Individual fish were prepared for human consumption in the manner used locally by Native Americans (headed, eviscerated, and scaled) and analyzed for Pb, Cd, and Zn. Whole crawfish were analyzed as composite samples. Metals concentrations were typically higher in samples from sites most heavily affected by mining and lowest in reference samples. Within the TSMD, most metal concentrations were higher at sites on the SR than on the NR and were typically highest in common carp and crawfish. Higher concentrations and greater risk were associated with fish and crawfish from heavily contaminated SR tributaries than the SR or NR mainstreams. Based on the results of this and previous studies, the human consumption of carp and crawfish could be restricted based on current criteria for Pb, Cd, and Zn. Overall, the wildlife assessment is consistent with previously reported biological effects attributed to metals from the TSMD. The results demonstrated the potential for adverse effects in fish, wildlife, and humans and indicate that further investigation of human health and ecological risks to include additional exposure pathways and endpoints is warranted.

A translocation of red swamp crawfish to different sites in the River Guadiamar in southwest Spain was performed to assess the ability of the species as a bio-indicator of heavy metal and metalloid pollution (Alcorlo et al, 2006). Crawfish were caged and exposed to a polluted environment for 6–12 days at three sites with different pollutant concentrations. Tissue (exoskeleton + gills, hepatopancreas, abdominal muscle) were dissected and analyzed by inductively coupled plasma-mass spectrometry (ICP-MS) to assess Cd, Cu, Zn, Pb, and As concentrations. Both exposure times resulted in significant bio-accumulation of some metals in crawfish tissue versus their environmental concentration. According to overall metal concentration, crawfish tissue ranked as follows: hepatopancreas/viscera >exoskeleton/ gills >abdominal muscle. Essential metals for crawfish metabolism (Cu and Zn) always occur in high concentrations independent of their environmental concentration due to the ability of crawfish to manipulate concentrations for their metabolic profit. Metals not involved in crawfish metabolism (Cd, Pb, As) tended to increase with increasing environmental concentrations and with longer exposure times. Thus, crawfish could be used as bio-indicator of these pollutants because their dose and time-dependent accumulation may be reflective of concentrations of nonessential metals in polluted wetlands. Future guidelines for plans to monitor pollution in Mediterranean rivers and wetlands should account for implementing crawfish incubation for 6 days and their subsequent metal content analyses as a routine.

In Spain a study completed using crawfish (*Procambarus clarkii*, males and females) that were exposed simultaneously to Cd and Zn during 21 days (Martin-Diaz et al, 2006). Exposure concentrations were those determined at the Guadiamar River after the Aznalcollar mining spill (SW, Spain): 10 and 30 mg/L of Cd and 1000 and 3000 mg/L of Zn. Three biomarkers (MT: metallothioneins like proteins, VTG: vitellogenin/vitellin like proteins and histopathology) together with heavy metal bio-accumulation were determined in soft tissues of male and female crawfish. At the concentrations tested, increasing Cd exposure resulted in increasing Cd bio-accumulation and increasing sublethal effects (induction of MT, VTG, and histopathology damage in tissues). Nevertheless, although increasing Zn exposure showed increasing VTG induction and histopathological damages, a positive relationship was not determined with MT induction. The only differences found between sexes were at the highest Cd exposure concentration related to bio-accumulation in hepatopancreas tissues. Biomarkers responses to heavy metal contamination in this crawfish, even VTG induction not before tested in heavy metal contamination assessment in crustaceans, resulted in potential tools for the monitoring of heavy metal environmental contamination.

Biomagnification of some essential metals (Fe, Zn, Cu) and toxic metals (Pb, Ni, Cd, Cr, Co, Mn) was determined in sediment, three types of fish (*Oreochromis niloticus, Synodonthis, Clarias gariepinus)*, and crawfish from the Ondo State coastal region, Nigeria (Asaolu and Olaofe, 2005). Metal biomagnifications in fish and crawfish was many times greater than in water; in sediment it was several thousand fold greater than in organisms and water. Among metals in water, Fe was the most abundant, with average concentrations of 146.7 and 74.3 mg/L in wet and dry seasons, respectively; Co was least abundant at concentrations of 2.4 and 1.6 mg/L. The Fe concentration was found to have an average concentration of 50.9 mg/kg in *C. gariepinus* and was the most abundant metal in fish; Cu, with an average concentration of 0.3 mg/kg in *O. niloticus*, was the least abundant metal. Biomagnification of most metals in both seasons varied widely from site to site. This was confirmed by a coefficient of variation from 31 to 144% and 29 to 130% in wet and dry seasons, respectively. Results showed that fish, crawfish, and sediment can be used to monitor metal pollution in Nigerian coastal seawater.

Water quality assessment in the Aznalcollar area of Spain was attempted using multivariate methods based on heavy metal concentrations in red swamp crawfish *(Procambarus clarkii)* (Sanchez Lopez et al, 2004). Trace levels of four heavy metals, Cu, Zn, Cd, and Pb, were detected in crawfish from 11 different stations. Principal component analysis (PCA) highlighted a gradient of contamination between the sampling stations. Cluster analysis (CA) distinguished three groups of stations. Discriminant analysis also differentiated three groups. The group centroids of the first discriminant function were used to devise an index that varies according to the source of the crawfish. These standardized values are proposed for use as a water quality index. The ability of this index to successfully predict environmental quality was proved with random samples.

Contamination of the American red crawfish from the Guadiamar riverside is due to the disastrous toxic spill that occurred on April 25, 1998, in the mining area of Aznalcollar-Los Frailes, Spain Sanchez Lopez et al 2003). A high concentration of heavy metals in the waters from the mine pool and their spill to the River Guadiamar was the cause of the destruction of a great number of animal and vegetable organisms. An inductively coupled plasma-mass spectrometry (ICP-MS) method for the total determination of heavy metals (Cu, Zn, Pb, and Cd) in whole bodies of American red crawfish *(Procambarus clarkii)* was used. Metals were extracted from the matrix in a closed-vessel microwave digestion system with nitric acid and hydrogen peroxide. A study of the uncertainty of the method for the determination of metals was carried out; at a concentration of 5 mg/L, the uncertainty was below 34%.

10.4 Research from McNeese State University in Lake Charles, Louisiana

Due to a national emergency in the early 1940`s, Southwest Louisiana and Lake Charles was chosen as a place for petrochemical refining and associated petrochemical industry. Currently there are about fifty different petrochemical companies in this area. Due to the lack of or enforced environmental laws and accidental spillage, the area was polluted , particularly with selected metals. While this has significantly improved since the 1970`s, there is still a legacy of contamination. Lake Charles is situated around 50 miles from the Gulf and can only be reached via a canal. This has to be continuously dredged which disturbs dormant or at least less accessible metal pollutant. Further complicating this area was a direct hit from Hurricane Rita in late September 2005 and to a lesser extent Hurricanes Gustav and Ike in September 2008.

Initial work in this laboratory at McNeese State University in Lake Charles, Louisiana was started back in 1997 when Dr. Joseph Sneddon was asked by a colleague, Dr. Mary G. Heagler from the Environmental Sciences Department to assist with the determination of lead using flame atomic absorption spectrometry with crawfish digestion via classical acid reflux in the meat of locally caught crawfish (Briggs-Reed and Heagler, 1998). The authors' data supported the hypothesis that the lead accumulated in the meat or digestive tract as an indicator that lead existed in the sediments of the soil in which the crawfish lived and ate. The concentration found ranged from 0.25 to 0.40 micrograms of lead per gram of crawfish. They recommended purging of the crawfish prior to human consumption to reduce the possibility of lead consumption.

Approximately ten years later this laboratory embarked on several major projects in the determination and interpretation of metals in crawfish. Initial work involved an undergraduate chemistry major project to determine copper, iron and zinc in the crawfish from Southwest Louisiana (Hagen and Sneddon, 2009). The results showed no significant differences in these metal concentrations between male and female crawfish, and no significant differences between crawfish from a pristine area and near a major highway. The concentrations of iron in the meat compared to the whole crawfish was about four times higher and was assumed to be due to the shell of the crawfish containing high levels of iron.

In early 2007 and through 2010, two major field trials were initiated (Richert and Sneddon, 2008b and Moss et al., 2010). Six metals (cadmium, copper, nickel, lead , iron and zinc) were determined by inductively coupled plasma-optical emission spectrometry (ICP-OES) in both tail (meat) and whole body of the crawfish and also in the soil in a four-month season (February through May) 2007 in Southwest Louisiana (Richert and Sneddon, 2008b). Cadmium or lead were not found in the meat or the while body. Nickel was found in some samples of the meat but mostly in the April and May samples. Copper, iron and zinc were found in both tail and whole body of the crawfish. Limited soil sampling showed no cadmium, nickel, or copper but levels or concentrations of iron, lead and zinc were found in the soils. Moss et al, (Moss et al., 2010) provided a mini-review of crawfish aquaculture and determined the following concentrations (microgram of metal per gram of dried sample of metal0, 95 % confidence interval and range from February through May 2009 as follows: cadmium, 0.49 +/- 0.14, 0.34-0.79, copper, 34.9 +/- 5.3, 23.8-44.2, nickel, 1.83 +/- 0.54, 1.08-3.39, lead, 18.0 +/- 4.0, 9.9-23, zinc , 47.4+/- 4.63, 41.3-55.8 were relatively constant with a slight increase in iron, 620.4 +/- 205.8, 328.8-1072.8 over the four months. The temperature of the crawfish ponds were monitored weekly but had no noticeable effect on the metal concentrations. Also sampled were the soil the crawfish lived in with copper and zinc concentrations decreasing with increasing water temperature and noticeable effect with the other four metals. A comparison to the study by Richert and Sneddon (Richert and Sneddon, 2008b) showed no significant differences in the six metal concentrations from one season to the next.

Two projects were undertake in the laboratory for controlled studies for the uptake of copper, lead and zinc in crawfish (Neelam et al, 2010) and selenium–lead interactions (White et al., 2012). Ten gallon aquarium tanks were spiked with up to 100 part per million (ppm) of the three metals and the crawfish introduced into the tanks. Not surprisingly all three metals were absorbed by the crawfish. Copper showed the highest absorption by the crawfish. Lead showed a constant increase in absorption with increased spiked of lead. It was noted that there was no obvious correlation between the metal absorbed and the amount of metal in the aquarium. It was noted that higher amount of lead were absorbed compared to copper and zinc. White et al. (White et al., 2012) determined whether the relationship between selenium and lead is one of an antagonistic or synergistic nature. Experiments were conducted on the freshwater crustacean crawfish (Procambarus clarkii). Crawfish were exposed to a known concentration of lead, dissected then analyzed to determine the accumulation of lead as follows: gills>exoskeleton>organs>edible meat Duplicates of the lead exposed crawfish were exposed to a concentration of approximately 10ppm (mg/L) selenium for a week to determine any adverse physiological effects. Within 48 hrs of exposure of selenium, the control crawfish were experiencing lethargy and signs of paralysis. The same symptoms began to occur to the previously lead exposed crawfish within 72 hours of selenium exposure. Analysis of the selenium exposed crawfish revealed time dependent and tissue specific adsorption of selenium identical to the concentrations of the lead exposed crawfish: gills>exoskeleton>organs>edible meat

After seven days of living in a complete state of paralysis, duplicates were placed in pure water to determine the ability of the crawfish to purge the selenium and regain mobility. Within 24 hours of purging, 88 percent of the paralyzed crawfish had regained full motor skills. Analysis of the purged crawfish showed a significant decrease in the concentration of selenium in the chitin rich exoskeleton and gills. However, the lead concentration in the gills and exoskeleton of the purged versus non-purged crawfish did not show any significant decrease indicating the covalent bond between the nitrogen and lead is much stronger than the chelating ionic bond between the selenium and lead.

Currently in progress is the use of crawfish shells for the uptake and removal of metal ions in aqueous solution (Vootla et al, 2011, and Beeram et al, 2011). Using ICP-OES, the samples were analyzed for various concentration of lead at three different volumes, 40-mL, 500-mL, and 3000-m.L. Crawfish shells , with the meat removed, were dried and pulverized to a 150-mesh size. The results showed that with an increase in volume of water, the capacity to remove lead by the crawfish exoskeleton by the same amount of shell powder decreased. However, lead absorption by the same amount of shell powder in all phases was good in terms of efficiency. The phase III study (3000-mL) showed that 0.5-g of crawfish absorbed the maximum amount of lead. Moreover, both raw and boiled crawfish shells have the same or similar capacity to uptake lead from water Vootla et al., 2011). Crawfish shell is a good source of chitin constituting about 23.5 percent on a dry basis. Using this value, the number of moles of chitin were calculated theoretically and compared to the number of moles of Pb that it can uptake. For 0.5 g of crawfish shell powder taken, the amount of Pb up take per gram of ground crawfish powder is calculated as shown in equation (1) and the values shown in equation (2) and equation (3)

$$\text{weight of crawfish} \times \frac{0.235 \text{ g of chitin}}{1 \text{ g crawfish}} \times \frac{1 \text{ mole chitin}}{\text{molar mass chitin monomer}} \times \frac{\text{molar mass of Pb}}{1 \text{ mole Pb Pb}} \times \frac{1 \text{ atom of Pb}}{1 \text{ monomer of chitin}} \quad (1)$$

$$0.5 \text{ g} \times \frac{0.235 \text{ g}}{1 \text{ g}} \times \frac{1 \text{ mole}}{203 \text{ g}} \times \frac{207 \text{ g Pb}}{1 \text{ mole}} \times \frac{1 \text{ atom Pb}}{1 \text{ monomer of chitin}} = 0.1198 \text{g Pb} \quad (2)$$

$$\frac{0.1198 \text{ g Pb}}{0.5 \text{ g of crawfish shell}} = 0.240 \text{g of Pb per 1g of crawfish shell} = 240,000 \text{ ppm Pb} \quad (3)$$

These calculations were obtained by assuming this model as 100 % chitin efficient crawfish shell. Based on 100% absorbance, one monomer uptaking one atom of Pb, the approximate value obtained was 240,000 ppm. The experimental value (amount of Pb that has up taken) obtained by analyzing the powder in this phase is approximately 200,000 ppm which is 83%.

Lack of 100% efficiency may be because some of the chitin molecules are inside the chunk of powder and not available for binding to the metals.

Work by Beeram et al., (Beeram et al., 2011) is still very much in progress as of early August 2011. This work concentrates on the use of the whole crawfish and shells (as opposed to ground crawfish shell by Vootla et al., (Vootla et al., 2011)) for the removal of various selected metals. Results are very preliminary but suggest surface area (ground versus whole) do play a part in the uptake rate.

11. Acknowledgements

JS acknowledges, with thanks, the support of the McNeese State University Alumina Research Award for 2010-2011 and 2010-2011 Shearman Grant fromMcNeese State University.

12. References

Alcorlo, P.; Otera, M.; Crehuet, M.; Baltans, A.; Montes, C.; (2006) The use of red swamp crawfish *(Procamburas clarkii)* as indicators of the bioavailablity of heavy metals in environmental poisoning in the River Guadimar (SW, Spain). *Science of the Total Environment*, 366(1), 380-390.

Asaolu, S.S.; Olaofe, O.; (2005) Biomagnification of some heavy and essential metals in sediments, fishes and crayfish from Ondo State coastal region, Nigeria. *Pakistan Journal of Scientific and Industrial Research*, 48 (2): 96–102.

Beeram, S.; Hardaway, C.J., Richert , J.C. Sneddon, J. (2011) Continued studies on the use of crawfish shells to remove selected metals ions from solution , *Analytical Letters*, in progress.

Briggs-Reed, L.M.; Heagler, M.G.; (1998) A comparative analysis of lead concentrations in purged and unpurged crayfish *(Procambrus clarkii)*: the significance of digestive tract removal prior to consumption by humans. *Microchemical Journal* 55 122-128.

Francuski, B.B., Milic-Djordevic, V.V., and Vojinovic-Miloradov, M.B. (2000) The level of pollution of the Tisa River by heavy metals and the cyanide after ecological catastrophe in the gold mine in Romania. Paper presented at the 5th International Symposium & Exhibition on Environmental Contamination in Central & Eastern Europe, Proceedings, Prague, Czech Republic, September 12–14; 2001.

Hagen, J.P.; Sneddon, J.; (2009) Determination of copper, iron, and zinc in crayfish (Procambrsu clarkia) by inductively coupled plasma optical emission spectrometry. *Spectroscopy Letters*, 42(1) 58-61.

Issacs, J. C.; Lavergne, D.; (2010) *Louisiana Commercial Crawfish Harvesters Survey Report.* Louisiana Department of Wildlife and Fisheries Socioeconomic Research and Development Section. March 2010.

Maranhao, P.; Marquies, J.; Madiera, V.; (1999) Zinc and Cadmium Concentrations in Soft Tissues of the Red Swamp Crayfish *Procamburas clarkii* (Girard, 1852) After Exposure to Zinc and Cadmium. *Environmental Toxicology and Chemistry*, 18(8), 1769-1771

Martin-Diaz, M.L., Tuberty, S.R., McKenney, C., Jr., Blasco, J., Sarasquete, C., and DelValls, T.A. (2006) The use of bioaccumulation, biomarkers and histopathology diseases in

Procambarus clarkii to establish bioavailability of Cd and Zn after a mining spill. Environmental Monitoring and Assessment, 116 (1–3): 169–184

McClain, W. R.; (2005) Crawfish culture in forage-based productions systems. American Fisheries Society Symposium. 46:151-169

McClain, W. R;, Romain, R. P.; Romaine, R. P;, Lutz, C.G.; Shirley, M.G.; (2007). *Louisiana Crawfish Production Manual*. Louisiana State University Agriculture Center and Louisiana Crawfish Production Board. Publication #2637.

Moss, J.C.; Hardaway, C.J.; Richert, J.C.; Sneddon, J. (2010)Determination of cadmium, copper, iron, nickel, lead and zinc in crawfish (*Procambrus clarkii*) by inductively couple plasma spectrometry: a study over the 2009 season in Southwest Louisiana, *Microchemical Journal* 95(10 5-10.

Neelam, V.; Hardaway, C.J.; Richert, J.C.; Sneddon, J.;)2010) A laboratory controlled study of uptake of copper, lead, and zinc in crawfish (*Procambrus clarkii*) by inductively coupled plasma optical emission spectrometry, *Analytical Letters*, 43 1770-1739.

Richert, J.C.; Sneddon, J.; (2008a) Determination of inorganic and organics in crawfish. *Applied Spectroscopy Reviews*, 43(1) 49-65.

Richert, J.C.; Sneddon, J. (2008b) Determination of heavy metals in crawfish (*Procambrus clarkii*) by inductively coupled plasma optical emission spectrometry: a study over the 2009 season in Southwest Louisiana, *Microchemical Journal*, 95(1) 5-10.

Sanchez Lopez, F.J.; Gil Garcia, M.D.; Sanchez Morito, N.P.; and Martinez Vidal, J.L. (2003) Determination of heavy metals in crayfish by ICP-MS with a microwave-assisted digestion treatment. *Ecotoxicology and Environmental Safety*, 54 (2): 223–228.

Sanchez Lopez, F.J.; Gil Garcia, M.D.; Martinez Vidal, J.L.;Aguilera, P.A.; Garrido Frenich, A.; (2004) Assessment of metal contamination in Donana National Park Spain) using crayfish (*Procamburas clarkii*). *Environmental Monitoring and Assessment*, 93(1-3), 17-29.

Schmitt, J.C.; Brumbaugh, W.G.; Linder, G.L., Hinck, J.E.; (2006) A sceening-level assessment of lead, cadmium and zinc in fish and crayfish from Northeast Oklahoma, USA. *Environmental Geochemistry and Health*, 28(5), 445-471.

Simon, O. and Boudou, A. (2001) Simultaneous experimental study of direct and direct plus trophic contamination of the crayfish Astacus astacus by inorganic mercury and methylmercury. *Environmental Toxicology and Chemistry*, 20 (6): 1206–1215.

Vootla, S.; Richert, J.C.; Hardaway, C.J.; Sneddon, J.; (2011) Investigation of crawfish (*Procambrus clarkii*) shells for uptake and removal of lead in water. *Analytical Letters*, 44(12): 2229-2243.

White, R.R.; Hardaway, C.J.; Richert, J.C.; Sneddon, (2012) Selenium-lead interactions in crawfish (*Procambrus clarkii*) in a controlled laboratory atmosphere. *Microchemical Journal*, submitted.

10

Applied Ecophysiology: An Integrative Form to Know How Culture Environment Modulates the Performance of Aquatic Species from an Energetic Point of View

Carlos Rosas[1], Cristina Pascual[1], Maite Mascaró[1],
Paulina Gebauer[2], Ana Farias[3,4], Kurt Paschke[3] and Iker Uriarte[3,4]
[1]*Unidad multidisciplinaria de Docencia e Investigación,*
Fac. de Ciencias, UNAM, Puerto de abrigo s/n, Sisal, Yucatán
[2]*Centro I-Mar. Universidad de los Lagos, Puerto Montt, Chile*
[3]*Instituto de Acuicultura, Universidad Austral*
de Chile, Sede Puerto Montt
[4]*CIEN Austral*
[1]*México*
[2,3,4]*Chile*

1. Introduction

Ecological energetics as a part of ecophysiology, appears to have grown out of the Age of enlightenment and the concerns of the physiocrats, a group of economists who believed that the wealth of nations was derived solely from the value of "land agriculture" or "land development." Their theories originated in France, were most popular during the second half of the 18th century. Physiocracy is perhaps the first well-developed theory of economics (Danbom, 1979). The Age of Enlightenment (or simply the Enlightenment) is the era in Western philosophy, intellectual, scientific and cultural life, centered upon the 18th century, in which reason was advocated as the primary source for legitimacy and authority. It is also known as the Age of Reason. The enlightenment was a movement of science and reason. Ecological energetic began in the works of Podolinksy in the late 1800s, and subsequently was developed by the Soviet ecologist Stanchinskii, the Austro-American Alfred James Lotka, and American limnologists, Raymond Lindeman and George Evelyn Hutchinson. It underwent substantial development by H.T. Odum and was applied by system ecologists, and radiation ecologists to understand how the forcing factors modulate the ecosystem interactions (Weiner, 2000). Currently ecophysiology is a discipline that, in aquaculture, have been widely used to establish how the environment modulates the performance of animals in order to obtain the highest amount of biomass in the shortest possible time and cost. The use of physiological capacities of organisms to obtain biomass has been one of the basic premises of the application of ecophysiological studies to the production of aquatic organisms.

2. General aspects of applied ecophysiology

The physiological ecologist seeks to understand the organism in relation to its environment. Defining the environment in aquaculture is relatively easy because it is, in many cases, a relatively closed system with the exception of marine cages or extensive culture. Three basic concepts are involved in applied ecophysiology (1) perception, (2) distribution on the culture environment and (3) the environment. Different animals perceive their surroundings in different ways, and to some extent, physiological and behavioural responses are dependent on perception. There is always a danger that we will superimpose our own human perceptions on other species and environments. Aquaculture is plenty of procedures that farmers have been translated from agriculture to aquaculture with not always success and in so doing fail to appreciate the interaction between organism and culture environment. The aquatic environment is completely different from the terrestrial environment and thus the perception of animals is so different from that we have. Water is more dense than the terrestrial environment provoking that organisms have developed a series of adaptations to perceive smells, spaces and physical and chemical conditions completely different from those of terrestrial animals.

Distribution on the culture environment take into account that individuals have limited tolerance ranges of temperature, salinity and other physical and chemical factors. Thus, is a truism that populations of organisms not be found in abundance beyond the tolerance regions of most individuals. Physiological tolerance and functional morphology go a long distance toward predicting where a particular organism will not occur, but they often give little indication of how well or what the organism will be doing, whether it will occur at all, within its tolerance limits. Applied to aquaculture, ecophysiology tries to establish the tolerance and resistance of species to environmental variables in order to provide the best culture conditions and how must be the environment to obtain the maximum scope for growth. Aquaculture environment is one of the topics of applied ecophysiology because the structure of the environment, interaction between individuals, water chemical and physical characteristics, type of food, etc., affect the physiology of aquatic animals enhancing or limiting animal performance, and at the end the biomass production. In this context, seed density, behavior characteristics that enhance or reduce cannibalism, turbidity, ecological characteristics of the culture systems that include meio-fauna and vegetation, are analyzed.

A physiological response represents the sum of all cellular and biochemical reactions as influenced by the environment or the animal itself. For this reason, organisms are capable of reflecting any environment condition even before the effects are observed in the population and community level. In ecological energetics an energy budget equation is defined as the sum of the energy from food ingested, which is divided into metabolizable, egested and excreted energy. This energy will varied according to the effects of different extrinsic and intrinsic factors and therefore it is important to calculate the cost of production in terms of growth when considering total aquaculture activity. Thus, the animal production (P) or growth is represented by the difference between the absorbed energy and the energy lost in respiration and excretion, taking age, sex, and body type into account. As with growth, other measures of biological productivity, such as work, egg production and body condition, will also affected by type of food or/and environmental characteristics. Thus, a satisfactory diet/environment in terms of the best efficiency will be needed in order to achieve optimal animal production.

Applied Ecophysiology: An Integrative Form to Know How Culture Environment Modulates
the Performance of Aquatic Species from an Energetic Point of View

183

This chapter will dedicated to show some examples that illustrate how applied ecophysiology have been solving some questions related with culture organisms and its consequences on production. In this chapter an integrative perspective related with immune condition is included taking into account that many of healthy problems observed on cultured animals are derived from immune problems provoked by inappropriate culture environments. Finally, some aspects related with experimental designs that must be considered in studying of applied ecophysiology are proposed. To exemplify, two species groups were formed: molluscs and crustaceans. Into the mollusks, examples with bivalves, gastropods and cephalopods are presented. Into the crustacean, examples with crabs and shrimp are used to illustrate how applied ecophysiology can improve aquaculture. In the case of fish a new encyclopaedia of fish physiology (from genome to environment) was published recently (Farrell, 2011). For that reason fish were not included in this chapter.

3. Molluscs ecophysiology

3.1 The case of bivalve *Argopecten purpuratus*

Bivalve filter feeders are capable of ingesting living and inert particles suspended in the water column that are responsible for the energetic input. In some species was demonstrated that detritus may contribute to the diet during periods in which the environmental offering of phytoplankton is unable to satisfy their energetic requirements. Although organic materials from discarded feed and faeces from salmon culture contributes to the diet of pectinids maintained in culture in southern Chile, phytoplankton is the main nutrient source for bivalves aquaculture (Farias & Uriarte, 2006). As pointed out, the shell growth and biochemical composition of larvae give clear indications about changes in the quality of the environment that are basic to determine how water nutrients and phytoplankton modulates the biomass production of bivalves (Ferreiro et al., 1990).

Studying pectinid *A. purpuratus* several ecophysiological parameters were investigated to find the best conditions to cultivate this bivalve species. In a first step biochemical composition of larvae and spat was used to evaluate the energy metabolism and the nutritional condition of hatchery reared bivalves (Farías et al., 1998). Adults from commercial long lines were stimulated to spawn using temperature shock and total lipids, soluble protein and total carbohydrates of eggs, larvae and spat were evaluated.

From that result was evident that lipids were the main source of energy while protein is deposited into the biomass. At the same time, an increment on carbohydrates was observed suggesting that the metabolic pathways related with transformation of lipids into glucose could be activated during larvae development and putting in evidence that quality of food used during larvae culture greatly influenced the storing of reserves and the survival of the resulting spat (Fig. 1).

These results were confirmed latter when the effect of dietary protein content on biochemical composition of postlarvae, spat and gonadal development of *A. purpuratus* was tested (Uriarte & Farías, 1999; Farías & Uriarte, 2001). From that results it was evident that pectinid, and other bivalves production depends on the quality of the diet and the environment in which they inhabit (Fig. 2). Results on reproductive conditioning of adults showed that feed quality affect the quality of eggs depending of the species, so for *A. purpuratus* an increase in micro algal protein produce an increase in fertility and growth rate

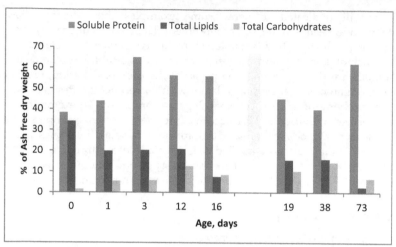

Fig. 1. Biochemical composition changes during larvae and postlarvae development of *A. purpuratus*. Arrow indicates metamorphosis.

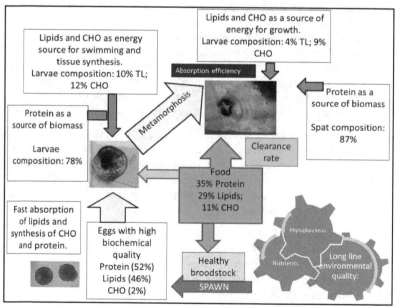

Fig. 2. Nutrient flows and biochemical pathways of planktonic larvae, spat and broodstock of *A. purpuratus* under culture condition. High quality eggs from healthy broodstock favors the use of lipids as a source of energy and synthesis of glycogen, while protein are used to protein deposition . After metamorphosis, carbohydrates and lipids are used mainly as a source of energy for maintenance and to accumulate protein via biomass production of spat. Food quality, measured as protein content, and environmental quality (Temperature, salinity, dissolved oxygen, turbidity etc.) determine the growth rate of bivalves, being the clearance rate and energy absorption key of culture success.

Applied Ecophysiology: An Integrative Form to Know How Culture Environment Modulates
the Performance of Aquatic Species from an Energetic Point of View

185

of larvae while in *Crassotrea gigas* non effect in fecundity was observed but eggs showed the highest lipid reserves and larvae showed highest survival. The scope for growth and reproductive conditioning of *A. purpuratus* is the product of the combined effect of high clearance rate and absorption efficiency that animals showed when fed high protein levels, indicating that phytoplankton abundance and composition will determine culture success (Farías & Uriarte, 2001).

The cost of the feeding depends as much on the processes pre- as post-ingestion and its modelling allows predicting, quantitatively, the growth of the bivalves based on simple but robust relationships between the conducts of feeding and the quality and amount of the food available in the environment. Although the importance of food quality and environment have been acknowledge, variability of the environmental quality and food remains as the principal concern between mussel and bivalve farmers, that depends on the natural food and environmental quality to maintain long line marine bivalves production (Lovatelli et al., 2007).

3.2 The case of gastropod *Haliotis fulgens* and *H. rufescens*

From the 130 species Halliotidae reported worldwide, only 14 species have been commercially exploited, either from fishery or through aquaculture. Since the commercial overfishing has severely depleted most populations of abalone, their aquaculture is replacing great part of the demand of these species. For abalone or any other cultured species, feed composition and ingestion rate are between most important factors to consider to predicting biomass production. In this sense, the study of *H. fulgens* have been used as an example to understand how the type of diet modulates the use and destination of the ingested energy, in this particular case, in attempt to optimize the use of macro algae or balanced foods.

Energy budget is a basic tool to evaluate the effect of type of food or other factors on physiological condition of aquatic animals. Using this model and considering different sizes of organisms, it is possible to obtain enough information to estimate their energy needs, and in a particular condition, predict the energetic costs associated with a particular environmental condition or type of food. In *H. fulgens* the energetic balance of animals between 0.1 to 2.5g was obtained in animals maintained in optimal conditions (Table 1) (Farías et al., 2003).

Relationship	Type of model	a	b	r^2
Dry body weight (g) DBW vs Total lenght (mm) L	$DBW = a\,L^b$	0.00001	3.08	0.98
Wet body weight (g) WBW vs Live weight (g) LW	$WBW = a\,LW\text{-}b$	0.15	0.63	0.99
Dry body weight (g) DBW vs Live weight (g) LW	$DBW = a\,LW\text{+}b$	0.02	0.11	0.98
Ingestion rate (mg day^{-1}) I vs dry body weight (g) DBW	$I = a\,DBW^b$	24.3	0.6	0.69
Absorption rate (mg day^{-1}) Ab vs dry body weight (g) DBW	$Ab = a\,DBW^b$	21.01	0.6	0.69
Respiration rate ml O_2 day^{-1} R vs dry body weight (g) DBW	$R = a\,DBW^b$	12.01	12	0.69
Nitrogen excretion (μmol N-NH$_4$ day^{-1}) U vs dry body weight (g) DBW	$U = a\,DBW^b$	43.6	0.85	0.63

Table 1. Relationship between morphometric characteristics and physiological responses of different sized green abalone *H. fulgens*.

With these equations it is possible to calculate how much of energy abalones needs per mg of biomass, how much of energy could be obtained, as a scope for growth per joule of ingested food (Table 2) and at the end the quantity of food that is needed to produce 1g of living abalone.

Physiological response		joules day^{-1} mg^{-1}	%
Ingestion rate	I	457	100
Feces	H	62	14
Absorption rate	Ab	395	
Nitrogen excretion rate	U	16	4
Respiration rate	R	172	38
Scope for growth	SFG	207	45

Table 2. Energetic balance obtained on optimal culture conditions of *H. fulgens*. Values obtained from equations derived from table 1. Circle indicates the proportion (%) of ingested energy (100%) that is channeled to each physiological response.

To make the calculations it is necessary to consider that 1g living abalone have 0.13g of dry weight biomass DBW (Table 1), that food have 18000 joules g^{-1} (Farias et al., 2003) and that scope for growth obtained was around 207 joules day^{-1} mg^{-1} DBW (Table 2). Following the next steps:

1. Joules day^{-1} needed to produce 1g living abalone =

$$\frac{\left(130 \text{ mg DBW of abalone contained into 1g living weight}\right) \times \left(207 \text{ joules day}^{-1}\text{mg}^{-1}\text{DBW}\right)}{1 \text{ mg DBW}}$$

$$= 35,100 \text{ joules day}^{-1}\text{g}^{-1}\text{LW}$$

2. If 1g of food have a content of 18,000 joules g^{-1}, but only 45% of it (8,100 joules g^{-1}) are really converted in biomass, then the quantity of food required to produce 1g of living abalone will be:

Food to produce 1g of living abalone = 35.1 kjoules day^{-1} g^{-1}/8.1 kjoules g^{-1} = 4.3g or 4.3kg of food for 1kg of abalone. With these models it is possible to help farmers to calculate not only biomass production levels but cost of that biomass production.

In a more generalized application of energetic balance results it is possible estimate the food needed to produce other abalone species as *H. rufescens* (Hernández et al., 2009). In that case formulated diet with 4.15 kjoules g^{-1} wet weight were compared with two macro algae with 2.7(*Porphyra columbiana*) and 0.92(*Macrocystis pyrifera*) kjoules g^{-1} wet weight energy contents, respectively. At the light of that energy contents and assuming the same conversion efficiency of *H. fulgens* (45%) it is possible to observe that to produce 1kg of abalone with macro algae will be necessary to use 4.15kg of formulated food, 28.9kg of *P. columbiana* and 84 kg *M. pyrifera*, both given as a fresh food. Results of the *H. rufescens* assay showed that abalones fed with macro algae grew 120% and 40% higher than observed in animals fed formulated diet suggesting that macro algae could be used to cultivate abalones instead of formulated diet. However natural production of macro algae

Applied Ecophysiology: An Integrative Form to Know How Culture Environment Modulates
the Performance of Aquatic Species from an Energetic Point of View

187

in high quantities that could be demanded to produce biomass could be a limit for the growth abalone culture.

3.3 The case of cephalopods *Octopus maya, Enteroctopus megalocyathus* and *O. vulgaris*

Of cephalopods, octopuses are considered economically interesting species for aquaculture. Landings in this area have been steadily decreasing since the 90's, which has led to increase the demand for octopuses and thus the technological-scientific efforts to rear them. Rearing is defined as the development of juveniles from an egg or paralarva with the ability to reach a second stage (Boletzky, 2003). Many cephalopods have been subject of several studies in captivity intended to investigate behavioural aspects (Hanlon & Wolterding, 1989; Hochner et al., 2006), used as models in neurophysiology studies (Flores, 1983; Wollesen et al., 2009), in predator-prey relationships (Villanueva, 1993; Scheel, 2002; Smith, 2003) or to provide live specimens for aquariums (Summer & McMahon, 1970; Bradley, 1974; Anderson & Wood, 2001). In the last years, many studies related with octopus and sepia culture have been putting in evidence the increased interest in their culture (for reviews see) (Sykes et al., 2006; Iglesias et al., 2007; Uriarte et al., 2011). Octopus are semelparous, that means that females spawn once in their life and die soon after the eggs hatch (Hanlon et al., 1991).The life span of the most of cephalopods species is short varying from about 6 months in small species to between one to three years in larger ones (Villanueva & Norman, 2008). Recent works show the very fast and high rates of food consumption, food conversion and growth of octopus. Most of our knowledge of octopus energy budget comes from studies of few species potentially due to their fisheries value and their potential as culture species.

Octopus vulgaris is by far the most studied between octopus species. At the date there are many investigations related with some aspects of energy budget. To exemplify the use of ecophysiological studies on octopus culture there are several studies in which octopuses were fed squid and tested at different experimental temperatures (Katsanevakis et al., 2005; Miliou et al., 2005). Katsanevakis et al. (2005) tested the effects of temperature and body weight on respiratory metabolism of juveniles of *O. vulgaris*. Using data obtained on that study a surface plot was constructed using a quadratic model for routine metabolism (Rrout) expressed as joules per day (Fig. 3).

That model was constructed using an interval of octopus living weight between 0.1 to 1kg and a sea water temperature between 13 to 28°C. With this information we can now calculate how much energy will be metabolized if octopuses of 0.1kg are cultivated at 15°C and in consequence how much food should be used to cover that energetic demand: Using the model:

$$Rrout = 19403.7 - [15.3\,(0.1kg)] - [2040.6(15°C)] - [0.003\,(0.1kg)^2] + [2.4\,(0.1kg \times 15°C)] + [52.2\,(15°C)^2] = 541.2\text{ joules day}^{-1}$$

To validate that model, we re-calculated the Rrout value obtained with last equation but using the living weight of octopuses used in other studies where Rrout was also measured. To do that we used weight values from García-Garrido et al. (2011) how made respirometric measurements of *O. vulgaris* (850g) fed squid at 15°C. The Rrout found by authors in that study was 671 joules day^{-1}, close to that obtained if the quadratic model is applied (554.4 joules day^{-1}). That concordance suggests that the quadratic model calculated now gave comparable data with other experimentally obtained (García-Garrido et al., 2011).

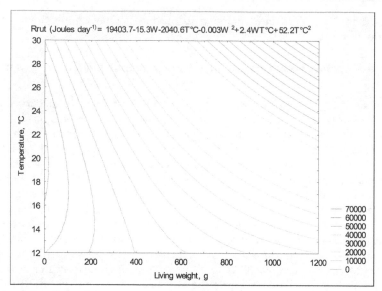

Fig. 3. Surface plot derived from a quadratic model to describe the relationship between Respiratory metabolism (joules day^{-1}) (Rrout = Routine metabolism; Rsda = Specific dynamic action metabolism; Rtot = Total metabolism) calculated for animals between 0.1 to 1kg living weight. (Data from Kasanevakis et al., 2005)

But, not always the models obtained give results applied at all mainly because the experimental procedures are not always are made in comparable conditions and depends on the form in which a determined factor affect the physiological response of animals. For example, using a model proposed by (Miliou et al., 2005) to evaluate the combined effect of temperature and living weight on feeding rate of *O. vulgaris* (Feeding rate g day^{-1} = -4.4608 + 0.9272lnW + 0.1138T°C - 0.0026T°C^2) we calculated the feeding rate (g day^{-1}) for octopuses fed squid with different living (W) weight and exposed at different temperatures (Table 3).

In their research, Miliou et al. (2005) feed octopuses with squid at a rate of 3% living weight day^{-1}. When the equation is applied the proportion of food consumed by octopuses resulted closed to that food rate (Table 3) suggesting that the data obtained with equation increased with temperature and living weight in a constant form (Table 3), producing a constant feeding rate. Using newly the study of García-Garrido et al., (2011), where octopuses were fed squid at a ratio of 10% living weight day^{-1} we observed that the model cannot predict the real feeding rate obtained by authors, that in this case was 8% living weight, because for octopuses feeding rate, and in consequence the energy derived of food, is modulated by the quantity of food that is offered, and certainly 3% could not be an ad libitum ratio.

Other species of octopus have been studied in attempt to quantify the physiological condition of octopus in similar conditions; a study was done comparing energetic balance of two species: *Enteroctopus megalociathus* (living at 10°C) and *O. maya* (living at 28°C) (Farias et al., 2009). The scope for growth of both species was calculated as: P = I - (R + H + U), and values of P of 522 and 358 joules day^{-1} g^{-1} were obtained of *E. megalociathus* and *O. maya*, respectively (Fig. 4).

Applied Ecophysiology: An Integrative Form to Know How Culture Environment Modulates
the Performance of Aquatic Species from an Energetic Point of View

189

Living weight,g	Temperature, °C	Feeding rate	
		g day^{-1}	% living weight
200	15	4.8	2.4
	20	5.4	2.7
	25	5.3	2.7
400	15	9.2	2.3
	20	10.3	2.6
	25	10.1	2.5
600	15	13.4	2.2
	20	15.0	2.5
	25	14.7	2.5
800	15	17.4	2.2
	20	19.6	2.4
	25	19.2	2.4
1000	15	21.5	2.1
	20	24.0	2.4
	25	23.7	2.4

Table 3. Relationship between living weight (g) and temperature (°C) on feeding rate of
Octopus vulgaris fed squid. Data obtained from the equation: Feeding rate g day^{-1} = -4.4608 +
0.9272lnW + 0.1138T°C – 0.0026T°C^2) (Miliou et al., 2005)

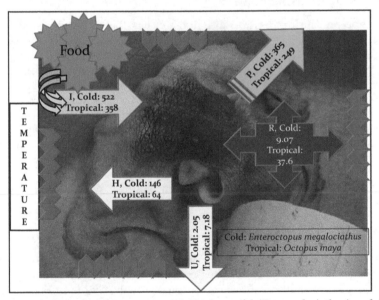

Fig. 4. The energy balance between species inhabiting cold (*E. megalociathus*) and tropical
environments *(O. maya)*, showed that the scope for growth (P) of *E. megalocyathus* was 1.5
times higher than that of *O. maya*. In that study was demonstrated that cold-water
cephalopod species could be more efficient than tropical species, a fact that should be
considered at the time of culture.

That study shows also that each octopod species is adapted to use ingested energy in a
different way as a consequence of the specific forces in each particular ecosystem: *O. maya*

loses more energy through ammonia excretion than *E. megalocyathus*, suggesting different physiological mechanisms for using the ingested protein (Table 4).

	a	b	R^2
Oxygen consumption, mgO$_2$ h^{-1} animal^{-1}			
Enteroctopus megalociathus	0.2	0.71	0.82
Octopus maya	0.93	0.69	0.93
Ammonia excretion, mgN-NH$_3$ h^{-1} animal^{-1}			
Enteroctopus megalociathus	0.32	0.37	0.53
Octopus maya	0.75	0.43	0.76
Ingestion rate, g day^{-1} animal^{-1}			
Enteroctopus megalociathus	0.19	0.67	0.95
Octopus maya	0.07	0.82	0.95

Table 4. Equation parameters ($Y = a\,W^b$) of oxygen consumption, ammonia excretion and ingestion rate of *Enteroctopus megalociathus* and *O. maya*. (Data from Farías et al., 2009).

4. Ecophysiology of crustacean

Between crustacean shrimp, crab and lobster has been deeply studied from an ecophysiological point of view. Crustaceans live in a wide variety of environments including fresh water, salty water and sea water. Crustaceans can be finding in holes in earth living with a little bit of humidity or in the crown of palms, several meters above the ground. Crustacean aquaculture has been growing in the last decades being the shrimp culture the most important activity. The world production of crustaceans was 5 million tons in 2008 being shrimp culture 73.3% of the total production (FAO, 2011). Today the white leg shrimp *Litopenaeus vannamei* is the most important species for aquaculture. *Litopenaeus vannamei* was introduced in China where its cultivation shifted to that of *Penaeus monodon*. Ecophysiological studies of shrimp have been important to define culture environments mainly when this species is mostly cultivated in salty water, where joint with diluted environment, animals are exposed to high temperatures, wide oxygen variations and nitrogen waste products.

4.1 Effect of salinity on shrimp physiological condition

White shrimp, *Litopenaeus vannamei*, the most cultivated shrimp species in America, grows well at salinities ranging from freshwater to hyper saline waters. Although *L. vannamei* tolerates broad salinity ranges, values for optimal growth have not yet been determined, and studies up to date show contradictory results. Whilst studies conducted under laboratory conditions have shown that maximal growth occurs between 5 and 15 UPS (Bray et al.(1994); Boyd, 1995; Rosas et al., 2001a), other studies have reported maximal growth at salinities above 35 UPS (Ponce-Palafox et al., 1997; Decamp et al., 2003). Recent studies have demonstrated a close relationship between nutrition and environmental characteristics at which shrimp are cultivated. This is because dietary levels of carbohydrates, lipids and proteins largely determine the shrimps capacity to respond to changes in the ionic composition of culture water.

Applied Ecophysiology: An Integrative Form to Know How Culture Environment Modulates
the Performance of Aquatic Species from an Energetic Point of View

191

4.1.1 Growth and survival

Salinity affects the distribution of a variety of estuarine and marine organisms. Some marine species, such as shrimps, have life cycles that include an estuarine phase. Since the 1950´s, there have been reports on congregations of juvenile shrimps at low salinities, whilst adults reproduced under strict marine conditions. It has also been observed that larval phases inhabit oceanic waters, where they develop and moult into postlarvae. Through certain chemo tactile mechanisms and water currents, postlarvae are capable of moving towards estuaries and coastal lagoons, where they recruit as early juveniles. It is in these habitats, with strong salinity variations, where some shrimp species are capable of colonizing more diluted environments than others. For example, studies carried out in the 1960´s by Mac Farland & Lee, (1963) showed that juvenile *L. setiferus* are better adapted to tolerate low salinities than juvenile *Farfantepenaeus aztecus*. Similarly, Mair, (1980) observed that amongst *L. stylirostris, F. californiensis,* and *L. vannamei,* the latter is the species that best tolerates highly diluted environments in estuaries and costal lagoons, and is therefore capable of inhabiting areas restricted to the other two species. This ecophysiological trait allows *L. vannamei* to reduce ecological pressures of competition for space and food, both strong limiting resources in these habitats. In an experimental study, Mair, (1980) observed that *L. vannamei* postlarvae and early juveniles placed in an experimental salinity gradient preferred salinities from 3-6 UPS, whereas *L. stylirostris* and *F. californiensis* preferred salinities from 32-35 UPS and 9-26 UPS, respectively.

The wide tolerance of *L. vannamei* to salinity has also been observed under culture conditions, both experimentally and within production facilities. Ogle et al. (1992) examined the effects of salinity on growth in *L. vannamei* postlarvae under different culture conditions. In that study the authors observed that tolerance to salinity is independent of postlarval (PL) age, showing that both PL8 and PL22 had better survival and growth at salinities ranging from 16 to 32 UPS. In that same study, it was demonstrated that tolerance to salinity increases as temperature declines from 30-16°C, since the highest survival and growth of postlarvae was obtained at salinities of 8 and 16 UPS at that temperature. Postlarvae can be acclimated to low salinities for production purposes. Recent studies have shown that a rate of salinity decline of 25% per hour is appropriate for postlarvae (PL10 to 4 UPS) and early juveniles (PL 20 to 1 UPS), demonstrating that this species is well adapted to highly diluted environments (McGraw et al., 2002). From these studies it is possible to conclude that *L. vannamei* postlarvae can tolerate broad salinity ranges, although shrimp between PL1 and PL10 are more sensible to low salinities that juveniles with more than 20 days after the last metamorphic moult (PL20).

In this study juvenile *L. vannamei* (initial mean weight 2.26g) were cultivated at 2, 4 and 8 UPS during 70 days in a semi-closed recirculating experimental system. Density was kept constant at 28 juveniles per m² in all salinity treatments. Results showed high survival rates in all treatments (98.7-100%). Mean final weight varied slightly between 19.0 -19.28g, whereas weekly growth rates varied between 1.67-1.7g. No significant differences in growth or survival amongst treatments were found, suggesting that *L. vannamei* can be cultivated at low salinities obtaining good survival and growth.

Bray et al., (1994) reported that juvenile *L. vannamei* positive to IHHN virus showed better growth exposed during 35 days at salinities of 5 and 15 UPS than at 25, 35 and 49 UPS. A weekly growth of 2g was obtained at low salinities 5 and 15 UPS. Mean final weights of shrimp were 12g (5 UPS), 12.2g (15 UPS), 10.8g (25 UPS), 11.1g (35 UPS) y 9,8g (49 UPS). No significant differences in mean final weight were found amongst low salinity treatments (5 and 15 UPS), but these were both significantly higher than that obtained at 49 UPS. Survival values were statistically similar amongst all treatments. A second group of juveniles with mean initial weights of 2.2g were infected with the IHHN virus, and maintained at 25 and 49 UPS. Their growth and survival was then compared with those of shrimp of the first group kept at those salinities. Results showed that growth of infected shrimp was similar to that obtained in the previous trial. In addition, there was a significant interaction between shrimp group and salinity, indicating that the group infected with IHHN grew less at high salinities. Mean final weights in the second trial were 10.3g (25 UPS) and 8.5g (49 UPS). Survival was also statistically similar for infected shrimp kept under the two salinity treatments.

In a more recent study Ponce-Palafox et al., (1997) reported that optimum growth and survival for juvenile *L. vannamei* were between 33 and 40 UPS and 28 and 30°C. Besides growth and survival of *L. vannamei* postlarvae were measured at 20, 25, 30, and 35°C and salinities of 20, 30, 35, 40 and 50 UPS. Groups of 30 individuals with three replicates were used in each temperature and salinity combination. Results clearly showed that juveniles of this specie survive best at 20 and 30°C and at salinities higher than 20 UPS. Shrimp growth was highest at temperatures between 25 and 35°C, with small differences amongst salinity treatments. Rosas et al.,(2001b) observed that better growth of juvenile *L. vannamei* can be attained when shrimp are kept at 15 UPS than at salinities of 35 UPS or more. Although much research on juvenile *L. vannamei* culture has been conducted, there is yet no decisive demonstration of the optimal salinity range. Whilst some researchers report that salinities lower than 25 UPS promote growth, others confirm that salinities higher than 35 UPS result best for this species. Combinations of salinity and temperature, as well as the type and quality of food appear to be determining factors to establish optimum salinity levels. Other shrimp species have also grown best at estuarine salinities

Chen et al., (1992) observed that *P. chinensis* kept at 25-31°C showed best growth and survival at salinities between 10 and 20 UPS. In addition (Staples & Heales, 1991) report best survival and growth in *P. merguiensis* kept under a combination of 20°C and 20 UPS. Rosas et al., (1999) showed that juvenile *L. vannamei* are energetically more efficient when exposed at salinities of 15 UPS than those lower (10 y 5 UPS) or higher (20, 25 y 30 UPS) than 15 UPS. In another study Brito et al., (2000) showed that juvenile *F. brasiliensis* have low tolerance to salinities lower than 25 UPS, below which shrimp stop growing and die. The authors explain that postlarvae and juveniles of this species occur in the southern Gulf of Mexico and Caribbean Sea, where they are rarely exposed to salinities lower than these. Rosas et al., (1999) found that salinity affects shrimp growth and survival differentially depending on postlarval age. These authors observed that optimal salinity for PL 10 lies between 30 and 35 UPS, whereas optimal salinity for PL 20 and older lies between 5 and 15 UPS. Optimal salinity ranges have been identified for other shrimp species. In general terms, species of the genus *Litopenaeus* are more tolerant to low salinities than those of the genus *Penaeus* or *Farfantepenaeus*. The former have optimal salinity ranges below 25 UPS, whilst the latter have optimal ranges at salinities above 30 UPS (Brito et al., 2000).

Applied Ecophysiology: An Integrative Form to Know How Culture Environment Modulates
the Performance of Aquatic Species from an Energetic Point of View

193

4.1.2 Relationship between salinity and food

Salinity is factor modulating the physiological state of shrimp. In costal environments where juveniles inhabit, estuaries and lagoons offer salinity conditions that fluctuate markedly both with tidal and seasonal variations. Penaeid shrimp have adapted to these changes taking maximum profit of food richness and protection found in these nursing areas. In order to survive and grow at the highest rates, shrimp have evolved through diverse physiological mechanisms that allow compensation for the marked changes in salinity. A decrease in water salinity results in a massive entry of water into the tissues by simple diffusion. In order to avoid extreme dilution and the consequent increment in cell volume, shrimp respond by: i) Reducing the concentration of ions dissolved in the cytoplasm. ii) Transporting amino acids from the cells to the blood, and then to the digestive gland. iii) Changing cell permeability at the gills. In crustaceans, ions transport has been extensively studied. Studies of Na, K and Cl transport as a result of exposure to diluted environments have shown that these ions play an active role in the reduction of cell volume by helping to keep osmotic pressure constant. However, because ions are small molecules with low osmotic power this mechanism is unsuccessful when shrimp are kept for long periods in diluted environments. A second mechanism has been described in which shrimp use the amino acid pool free in the cytoplasm to regulate cell volume. Amino acids are the principal component of proteins, and are abundant within shrimp cells. When a salinity change occurs, amino acids are transported towards the hemolymph where they are re-absorbed by the digestive gland. There, amino acids are stored as hemocyanin, as peptides for growth, or as proteins for the immune system. A recent study by Rosas et al.,(2002) includes a summary of the effects of dietary and protein and carbohydrate levels on the physiological and nutritional state of shrimp maintained at high and low salinities (Fig. 5).

It can be seen that CHO (carbohydrates) enter as starch and are transformed into glucose and glycogen. Laboratory studies have shown that starch breakup by amylase has a saturation point, indicating that shrimp capacity to breakup starch increases to a maximum level. At CHO dietary concentrations higher than 35%, the processing rate of amylase stays constant even with increasing CHO levels Rosas et al., 2000). Because in diluted environments amylase is less efficient, the saturation point of amylase has a larger effect under low than high salinities, (Rosas et al., 2001a). Another important aspect is the saturation of glycogen in the digestive gland (digestive gland) of shrimp fed with diets containing excessive CHO. Results of laboratory studies have shown that shrimp capacity to store glycogen is limited, and the digestive gland is rapidly saturated with levels of dietary inclusion higher than 30 %. When shrimp are kept at low salinities, this saturation has a larger effect, because it prevents adequate absorption of amino acids used to maintain internal osmotic pressure and cell volume Rosas et al., (2001a). These results have been interpreted taking into account that shrimp basically use CHO to form chitin, which in turn is the principal component of the exoskeleton. A glucose molecule and one ammonia radical product of the amino acid degradation form chitin. The combination of these products emphasizes the fact that chitin formation depends on CHO metabolism, which constitutes the center of penaeid shrimp nutrition.

The adaptations of shrimp cultivated in hypersaline environments have not yet been studied in depth. Studies conducted in our laboratory have shown that at high salinity the effects of dietary proteins and CHO operate in a different way than when at low salinity Pascual et al., (2003); Pascual et al., (2004a). From our results, it can be concluded that at high salinity

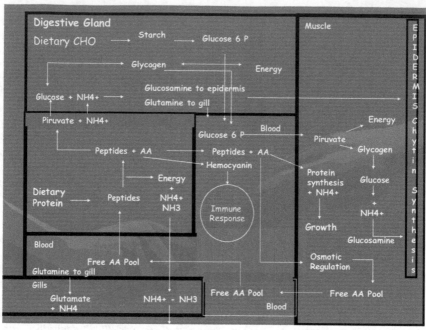

Fig. 5. Diagram showing a summary of the relationship between dietary proteins and carbohydrates and the metabolic pathways involved in the uptake of ingested nutrients.

shrimp may require lower protein levels than those they require at low salinity. This is because at concentrated environments ammonia produced by the degradation of dietary amino acids may be toxic. At low salinity degraded amino acids are transformed to ammonia and used for ion exchange, whilst at concentrated environments ammonia molecules are trapped in the blood and cannot be released as rapidly and efficient. Notwithstanding the high tolerance of crustaceans to high ammonia in blood Schmitt & Uglow, (1997), levels above 12 mg l-1 could inhibit growth, delay moult and reduce breathing capacity of shrimp Danford et al.(2001). High levels such as those can easily be attained when the diet contains more than 35% proteins and shrimp are kept at salinities higher than 40 UPS Rosas et al., (2000)

The effect of salinity on lipid metabolism in has neither been studied in depth. In a recent study the effect of salinity on shrimp survival, the activity of the Na/K bomb and lipids was examined Palacios et al., (2004). In that study 20 day old fed and unfed (3 and 24 h fasting) postlarvae were compared to demonstrate the effect of energy reserves on survival and osmoregulatory mechanisms. Activity of the Na^+-K^+ ATPase was 5 times higher in posterior gills than in anterior ones, a response associated with the osmoregulatory function of this organ. In addition, a non-significant increase in anhydrase activity was observed in postlarvae exposed for 20 days to 10 UPS. This is explained by the role of this enzyme in the hydration of CO_2 produced in respiration. Under low salinity conditions, the internal media dilutes and the output of Na + decreases. Na^+ in then exchanged for H^+ and Cl^- for HCO_3, provided by CO_2 through the carbonic anhydrase. In that same study total lipids in digestive gland were significantly lower in fasting than in fed PL20, whereas triacyl glycerids

Applied Ecophysiology: An Integrative Form to Know How Culture Environment Modulates
the Performance of Aquatic Species from an Energetic Point of View

195

were statistically similar. This is probably due lipids in the digestive gland being used during the first 3 hours of fasting, allowing for animals to have sufficient energetic reserves to tackle the salinity change Rosas et al., (1995). This study concluded that movement of lipids to satisfy the energetic demand of osmoregulation could be operating in shrimp of this species.

4.2 Physiological condition and immune system of shrimp

The shrimp immune system has a solid protein base and hemocyanin plays an important role in its function. Recent studies have demonstrated that in addition to its multifunctional role (oxygen transporter, storage protein, carotenoids carrier, osmolite, ecdysone transporter) hemocyanin has a fungistatic Destoumieux et al.,(2001) and prophenol oxydase-like function Adachi et al., (2003). Proteins are also involved in recognizing foreign glucans through lipopolysaccharide binding protein (LPSBP) and β glucan binding protein (BGBP) Destoumieux et al., (2000); Vargas-Albores & Yepiz-Plascencia, (2000). A clotting protein (with the change of fibrinogens to fibrin) is involved in engulfing foreign invading organisms and prevents blood loss upon wounding Hall et al., (1999); Montaño-Pérez et al., (1999). Defense reactions in shrimp are often accompanied by melanization. Prophenoloxidase (ProPO)-activating system, mediated by hemocytes, is a zymogen of phenoloxidase (PO) enzyme that catalyzes both o-hydroxylation of monophenols and oxidation of phenols to quinones leading to synthesis of melanin Sritunyalucksana & Söderhall, (2000). Conversion of ProPO to PO occurs through a serine protease called prophenoloxidase-activating enzyme (ppA) regulated by another protein, a-2 macroglobulin, a trypsin inhibitor Perazzolo & Barracco, (1997). The innate immune response of shrimp also relies upon a production, in hemocytes, of antimicrobial peptides called peneidins that are active against a large range of pathogens essentially directed against Gram-positive bacteria via a strain-specific inhibition mechanism Destoumieux et al., (2000).

In order to reach effects of dietary protein level on energetic balance we probe two protein levels in a range of optimal reported levels of 15% and 40% (equivalent to 15 and 40g DP/kg body weight/day [g DP/kg BWd]) and one extremely low (5% equivalent to 5g DP/kg BWd) were used to feed juveniles for 50 days. Dietary protein level enhanced ingestion rate in shrimp fed 5g DP/kg BWd compared to shrimp fed 40g DP/ kg BWd, however, daily growth coefficient (DGC,%) of L. vannamei juveniles was high in shrimp fed 40g DP/kg BWd. An inverse relation between wastes (H+U) and dietary protein level was observed indicating that shrimp lose 81% of ingested energy when fed 5g DP/kg BWd and only 5.6% when fed 40g DP/kg BWd. A higher assimilation and production efficiency (P/As) was obtained when shrimp were fed 40g DP/kg BWd, than obtained in shrimp fed 15 or 5g DP/kg BWd. An increase in Oxy hemocyanin was observed with increasing dietary protein levels indicating that shrimp accumulated protein as hemocyanin. A reduction of hemocytes occurred when shrimp were fed sub-optimal dietary protein; same patron was observed in the respiratory burst. The compensatory mechanism used by L. vannamei to respond nutritional stress, sub-optimal dietary protein level (5 and 15g DP/kg BWd) induced not only a severe reduction in growth rate and assimilation efficiency but also in immune capacities Pascual et al., (2004).

In an attempt to know how the protein level modulates catabolism and its effects on the immune response, we studied juvenile L. vannamei that had been starved for varying period after being conditioned on diet containing either maintenance or optimal dietary protein levels Pascual et al.,(2006). Juvenile shrimp were fed for 21 days on diets containing 5% and 40% dietary protein. Hemolymph metabolites (glucose, cholesterol, protein, acylglycerols, and lactate), hemocyanin, osmoregulatory capacity, digestive gland glycogen and lipids, and immune conditions (hemocytes concentration, phenoloxidase activity, respiratory burst: basal and activated) were evaluated and considered as initial condition. After that time, shrimp were starved for 21 days. A reduction in all physiological and immunological indicators was observed with starvation. The results demonstrate that shrimp are well adapted to tolerate food deprivation for some time but that this tolerance is closely related to its previous nutritional condition. In the case of shrimp fed 40% DPL, wet weight, nutritional and immune condition was significantly affected after 14 days of starvation. In shrimp previously fed 5% DP, tolerance to starving condition was limited to only a few days (7 days) as a result of low reserves of circulatory and mussel proteins. All these results demonstrate that dietary protein levels can governor the immune condition of shrimp through the management reserves metabolism.

Domestication is other important aspect to obtain better results during culture; however, artificial selection has important implication on physiological adaptations. Pascual et al., (2004) studied wild and seventh-generation cultivated shrimp to determine how size-based selection could alter the nutritional and immunological conditions of Litopenaeus vannamei. Wild juveniles and a sample of seventh-generation cultured shrimp were acclimated under identical conditions. During 55 days, shrimp were fed a high (HCHO: 44%) or a low (LCHO: 3%) carbohydrate diet. Wild shrimp showed a direct relation between dietary CHO and lactate, protein and hemocyte levels indicating that dietary CHO was used for protein synthesis via transamination pathways. In seventh-generation cultured shrimp these parameters were inversely proportional to dietary CHO level, indicating the capacity to synthesize protein from dietary CHO was repressed in cultured shrimp. Farmed shrimp showed a limited capacity to respond to LCHO diets demonstrating high protein dependence in their metabolism and immune response. These results demonstrate that during size-based breeding programs other metabolic process than CHO catabolism can be selected. The incapacity of shrimp to use dietary CHO could limit protein reduction of diets and limit the efforts of the shrimp industry to be ecologically and environmentally profitable.

Considering the results arising from a series of studies, Arena et al., (2003); Pascual et al., (2004 a and b), developed a conceptual model about how dietary components modulate the fate of energy intake, the nutritional status and immune system of juvenile L. vannamei (Fig. 6). The environment and genetic variability are the basis of the model, since as mentioned; the interaction between these elements affects the digestive capacity, the flow of energy, protein synthesis, osmoregulation, disease resistance and the degree homeostatic control.

Degradation of starch leads to glucose uptake, it is transported directly to the hemolymph or metabolized through glycolysis (**GL**); subsequently pyruvate through acetyl-coenzyme-A (**A-CoA**) can enter the Krebs cycle (**KC**) and continue with the respiratory chain (**RC**) to obtain energy (**ATP**). The digestive capacity of wild shrimp revealed the importance of taking advantage of dietary CHO, which can be associated with the various points of metabolic regulation of the glycolytic pathway. Nine of the ten essential amino acids can be synthesized from glucose, where glutamate provides the amino group. In the opposite

Applied Ecophysiology: An Integrative Form to Know How Culture Environment Modulates
the Performance of Aquatic Species from an Energetic Point of View

197

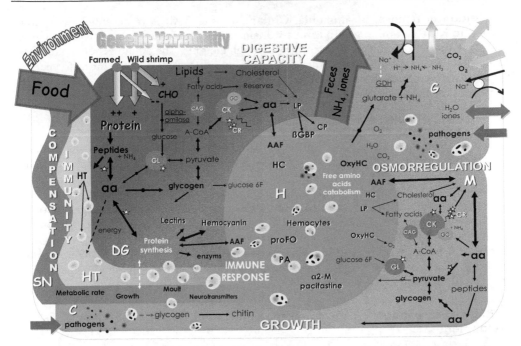

Fig. 6. Conceptual model: protein metabolism and homeostasis. The scheme includes the
main organs and systems of juvenile shrimp: digestive gland, **GD**, gills, **G**; muscle, **M**;
cuticle, **C**; nervous system, **SN**, hematopoietic tissue, **HT** and hemolymph, **H**. Genetic
variability is the basis of the model as it affects the flow of energy from food chemistry to
physiological processes. *Litopenaeus vannamei* juveniles grown in closed loop and sorted by
size (F7, F25) have a limited capacity to respond to carbohydrate diets (CHO) demonstrating
high protein dependence in their metabolism. The degradation of proteins generates
peptides and amino acids (**aa**) that are oxidized or used to synthesize **glycogen**. The aa can
also be used for the synthesis of digestive and metabolic enzymes, and immune proteins
(such as prophenoloxidase, **ProPO**, lectins, **L**, lipoprotein recognition, **BGBP**, clothing
protein, **CP**, antimicrobial peptides, **AP**; of regulation of the immune response; pacifastine
and α2-macroglobulin, and hemocyanin, **H**, which seems to play a role in the immune
response of shrimp, in addition to functioning as a storage protein in the hemolymph.
Amino acids can also be transported to the hemolymph to form the pool of free amino acids
(**FAA**), which can be used to generate muscle tissue.

direction, for example in conditions of nutritional stress, catabolism of glucogenic amino
acids is the basis for the gluconeogenic route for CHO can be used as energy substrate. The
cycle of fatty acids (**CAG**) is connected to the glycolytic pathway through **A-CoA**, a high
concentration of **ATP** or **NADH** promotes the synthesis of fatty acids from the A-CoA. The
ß-oxidation of lipids from food or reserves represents the reverse direction to generate
metabolic energy.

Regulation of osmotic pressure is mainly associated with the Na-K ATPase, ion exchange,
the catabolism of aa and the activity of glutamate dehydrogenase (**GDH**), which controls

the addition or removal of ammonia. Glutamate cycle (**GC**) is represented by a circle attached to KC. The amino group can be transferred to glutamine, which is less toxic than NH_4, so is involved in nitrogen excretion through the gills. The amino group can also be coupled to a carbohydrate to the formation of glucosamine for chitin formation. Therefore, growth and immunity of juveniles of *L. vannamei* are linked to domestication and protein metabolism.

The hematopoietic tissue and hemocytes are represented in the model under a deep interaction with the GD and the nervous system, since nutritional status and some hormones affect the rate of proliferation of hemocytes and thus the state of the immune response, because in the hemocytes are synthesized many of the immune effectors. On the other hand, the concentration of hemocytes in the epidermis during some stages of moult may be related to the release of glycogen for the synthesis of chitin, which is consistent with the ability of hemocytes to synthesize glycogen and store huge amounts (proportionately greater than content in the digestive gland). Also, the high amount of amino acids within the hemocytes could point to a possible release into the hemolymph associated with osmotic regulation and/ or energy demand Claybrook, (1983). Since the metabolic functions of hemocytes have not been demonstrated, in the model are marked with a dashed arrow.

4.3 Effect of temperature on energy losses during southern king crab larval development

Temperature is one of the most important environmental factors affecting every aspect of an organism's physiology, from the basic structures of the macromolecules that are responsible for catalysis and information processing to the rates at which chemical reactions occur Hochachka & Somero, (2002). Thus, energy acquisition, losses, allocation to maintenance and growth in any organism is affected by temperature. In heterotrophic organisms, the ingested food provides the energy input which is balanced with the main sources of energy losses i.e. bio deposition or faeces rejection, metabolism quantifiable by the oxygen consumption rate, and in less extent nitrogen excretion. The reminder energy is canalized to growth and/or reproduction. In a lecithotrophic (i.e. food independent) condition, the principal energy losses are related to respiration and nitrogen excretion.

The lithodids, commonly known as king crabs or stone crabs, inhabits high latitude cold waters of both hemispheres. Several species of king crabs represent valuable fisheries but commercial exploitation and environmental factors has dramatically declined the landings. In the southernmost part of South America (Chile and Argentina) *Lithodes santolla* (Molina), the southern king crab, is commercially exploited and severe fishing restrictions are necessary to protect the populations. Both circumstances increased the interest and expectative about the development of cultivation technologies for the southern king crab.

In order to establish larval rearing conditions, ecophysiological studies on *L. santolla* larvae have been carried out to define culture temperature based in the energy losses.

Early life-history stages tolerate low temperatures and their larval development is fully independent of food (lecithotrophic) Anger et al., (2004); Calcagno et al., (2004); Kattner et al., (2003); Lovrich et al., (2003); Thatje et al., (2003).

Applied Ecophysiology: An Integrative Form to Know How Culture Environment Modulates
the Performance of Aquatic Species from an Energetic Point of View

199

Our results showed that the development time fluctuate between 55 and 42 days approximately, at 9 and 15°C, respectively. The sustained even increasing respiration rate throughout the zoeal stages is interpreted as a confirmation of lecithotrophic development, otherwise a drop in metabolic rate in starved larvae were expected (Fig. 7a). Megalopa showed at lower temperatures a decreasing respiration rate attributed to a behavioral change in the swimming activity, while at higher temperature the metabolic rate increased (Fig. 7a). A similar pattern was observed for the nitrogen excretion rate (Fig. 7b). The atomic proportion of consumed oxygen and excreted nitrogen (O:N) is considered as an estimator of the metabolized substratum; values between 3 and 16 would indicate protein metabolism, an O:N ratio between 50 and 60 corresponds to similar proportion of lipid and protein catabolism and higher values than 60 represents lipid metabolism Mayzaud & Conover, (1988). *L. santolla* larvae reared at 12°C, in general showed a combined protein and lipid catabolism, but, at 9 and 15°C the Zoea I showed an evident lipid metabolism (Fig. 7c). Respiration rate, independent to temperature, contributed principally on the energy losses, ranging between 93 and 96% at 9 and 15°C. Although the shortest development time was at 15°C, low survival suggests unsuitable rearing conditions. Indeed, the highest energy losses expressed as excretion and respiration rate (Figs. 7 a and b) occurred at 15°C. Multiplying the energy losses of respiration and excretion rates by the respective developmental time, the cumulative energy losses were estimated. Similar energy losses during Zoea I and II were observed for the different temperatures, while an evident impact of higher temperature on the megalopa stage was observed (Fig 7d). The results suggest that elevated temperature ≥15°C could generate a mismatch on the larval metabolism of *L. santolla* due the increasing energy losses and the restriction in the limited stored energy in the form of yolk remaining

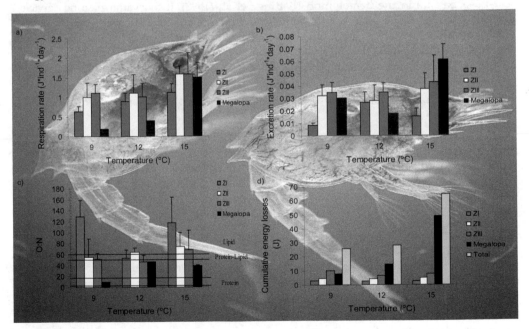

Fig. 7. Temperature effect on a) the respiration rate; b) the excretion rate ($J*ind^{-1}*day^{-1}$); c) O:N ratio; d) the cumulative energy losses (J) during larval development of *Lithodes santolla*.

from the embryonic development. For cultivation purposes physiological data suggest rearing temperatures below 15°C and special caution to reducing larval stress and wasteful energy losses.

After the increase in the atmospheric oxygen availability 2000 million of years ago, most animals were selected for a more efficient energy acquiring mechanism using oxygen as the last electron acceptor. While anaerobic pathways persist, aerobic ATP yield increase as much as 15 to 18-folds (depending on the metabolic pathway) the efficiency in the process of gaining energy Wieser, (1986). Hence, oxygen has a main role in obtaining energy for the cells, therefore for the organism. Oxygen decrease (hypoxia, both environmental and cellular; hypoxic generally refers to low oxygen conditions that are physiologically stressful, Levin, (2003)) has many effects in different levels, from population to individual and cellular, affecting the capability to obtain energy, which affects also the organism's activity and their growth performances. One of the most important mechanisms to deal with hypoxia is to improve the use of energy available, by means of reducing the metabolic rate. In fact, organisms consider well adapted to hypoxia events show a lower ATP consumption, to balance the reduction in the capability to generate ATP. During hypoxia, the ion pump slightly decreases, but the most important is the low in protein synthesis and degradation Hochachka & Somero, (2002). The reduction in protein synthesis result in an efficient energy saving mechanism and, at the same time, an effective mechanism to face hypoxic events, but for aquaculture purposes, such mechanisms activated by low dissolved oxygen (DO) results in a drop in the ingestion rate, animal activity and finally a marked reduction or even ceasing in growth. In marine animals, adaptation to low oxygen availability includes also modifying the acid-base balance in the hemolymph Martinez et al., (1998), hemocyanin binding capacity, oxyhemocyanin protein relationship, hemolymph osmolality, and ion concentrations as well as hemocyanin synthesis Johnson & Uglow, (1985); Charmantier et al., 1994; Chen & Kou, (1998); Taylor & Anstiss, (1999) Environmental parameters like high pH and low dissolved oxygen (DO) have been reported to cause reduction in hemocyte counts Cheng & Chen, (2000).

4.4 Dissolved oxygen as key factor in the bioenergetics of southern king crab juveniles

Crustacean aquaculture is widespread in tropical areas where higher water temperatures reduces oxygen solubility and accelerates the decomposition of organic material by bacteria and other oxygen consuming microorganisms both conducting to a decreasing in DO levels Rosas et al., (1998, 1999). For cold water organisms the relationship between DO and temperature described by Pörtner (2005) by the Metabolic Cold Adaptation hypothesis implies that the thermal tolerance of the animal is affected by the supply of oxygen to their mitochondria.

In an attempt to establishing the impact of culture environmental factors on the physiology of southern king crab Lithodes santolla with a final goal to develop cultivation techniques, the effect of dissolved oxygen (DO) on different physiological responses of L. santolla juveniles is analysed. Paschke et al. (2010) showed physiological responses in juveniles of L. santolla exposed 10 days to reduced Oxygen availability treatments (Fig. 8).

The juveniles regulate their metabolism until 4-9 kPa (ca. 4-6 mg O2*l⁻¹), maintaining their oxygen consumption rates, hemolymphatic protein levels, as well as the oxyhemocyanin

Applied Ecophysiology: An Integrative Form to Know How Culture Environment Modulates
the Performance of Aquatic Species from an Energetic Point of View

201

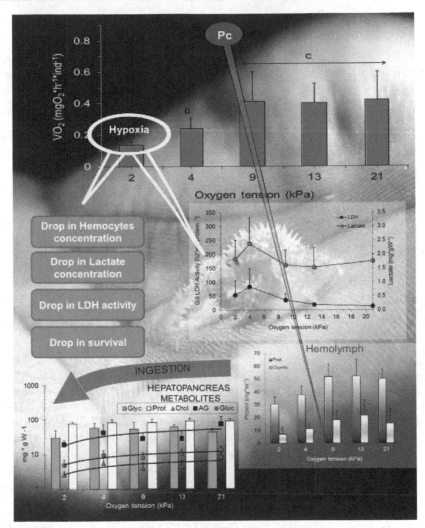

Fig. 8. Effect of dissolved oxygen (DO) on different physiological responses of *L. santolla* juveniles. Pc: critical point of aerobic metabolism shift; LDH: Lactate dehydrogenase; Prot: protein concentration; Oxy-Hc: oxygenated hemocyanin; Glyc: glycogen; Chol: cholesterol; AG: acylglycerids; Gluc: glucose.

concentration. These responses to reduced DO were complemented with anaerobic metabolism evidenced by the increase in the activity of gill-LDH and a consequent increase of Lactate concentration. A reduction in nutritional compounds in the hepatopancreas was interpreted as a consequence of a notorious reduction of food ingestion, and as an additional consequence of hypoxic conditions (the ingestion was affected severely and showed a linear relationship to dissolved oxygen (DO) between 2 and 13 kPa: 0.59 and 1.92 mg Dry weight h-1 ind-1, respectively). Severe chronic hypoxia (1 mgO2*l-1 or 2 kPa for 10 days) led to an unsustainable reduction in the aerobic and anaerobic activity, where a reduction of the gill-

LDH activity and Lactate concentration as well as oxygen consumption coincide with a significant reduction in hemolymphatic hemocytes and finally a survival reduction to 70%.

In accordance with these results and applying Ecophysiology to the development of cultivation techniques, the rearing systems were modified to ensure the adequate Oxygen supply in the first millimeter over the bottom surface, the place where juveniles of the southern king crab *Lithodes santolla* feed, moult and growth.

5. Modelling individual growth: Size-at-age variability in cephalopods and mixed models tools

Both individual and population growth is central to aquaculture practice and research, and modelling growth has been an important aim in methodological and theoretical studies Moltschaniwskyj, (2004); Vigliola & Meekan, (2009). Models have been extensively developed and applied to understand population growth (e.g. Malthussian models, logistic models, Leslie matrix, etc.), and those for individual growth have followed closely (von Bertalanffy model). Methods for measuring and modelling individual growth can be classified as either indirect or direct Semmens et al.,(2004). Indirect methods are those using modal progression analysis based on length-frequency data obtained from animals caught in the wild. These have been used to predict length-at-age in individuals of a variety of both vertebrate and invertebrate fisheries Solis-Ramírez, (1997); Cortez et al., (1999) Direct methods, such as mark and recapture using a variety of tags, have been used to follow the growth of juvenile and adult stages in the wild Sousa Reis & Fernándes, (2002); Cortez et al., (1999).

From an ecophysiological point of view, however, more useful are direct methods examining the growth of individuals whose age and specific culture conditions are known. The growth of an individual is the result of a series of energy transformations that begin with food ingested, together with the balance between the uses and destinations of the energy contained in that food Lucas, (1993); Rosas et al., (2007). Thus, the growth curve represents the manner in which the physiological and energy demands are expressed over time O'Dor & Wellls, (1987); Pauly, (1998) at differing levels of biological organization, i.e. body size, organs, tissue, and cells Moltschaniwskyj, (2004). In the context of energy balance, the ultimate need to measure and compare the way in which individuals grow becomes clear: individual growth models can be used not only to predict size at age, hence biomass productions at certain points in time, but can also serve to estimate growth curve parameters that can later be compared amongst different culture conditions.

The mathematical tool generally used to obtain growth curve parameters is least square fitting, which can be used both for linear and non-linear models. The linear least-squares problem occurs in statistical regression analysis where it has a closed-form solution. The non-linear problem has no closed-form solution and is usually solved by iterative refinement, where chi-square (X^2) and probability (p) values are obtained and evaluated at each iteration.

Analyses of individual growth have commonly involved fitting linear regression models of weight against time. However, such an approach involves difficulties because the weight vs. time relationship is rarely linear and, when it is, it is only for very short and specific periods. Moreover, repeated measurement of the same individuals violates an indispensible

Applied Ecophysiology: An Integrative Form to Know How Culture Environment Modulates
the Performance of Aquatic Species from an Energetic Point of View

203

requirement of regression models, namely that each data point be independent from others
Zar, (1999); Zuur et al., (2007). Finally, size-at-age variation generally increases with age,
resulting in strong heterogeneity of variance in different ages, another impediment for
ordinary regression analysis Zuur et al., (2007). These characteristics violate important
regression assumptions, thus producing unreliable models, i.e. models with dubious F, X^2
and p values that cannot be used for prediction purposes.

A methodological approach is fitting a regression model to the weight of each individual at
a known fixed time (t_2) against its weight at an earlier point in time (t_1). This method ensures
linearity of the X–Y function and independence of data points (as long as they are
individually labelled). The resulting linear equation describes a type of relative growth, and
its slope represents the proportionality of the difference between two individuals at t_2
relative to the difference between them at t_1. In turn, the line's intercept represents the final
weight reached by the smallest individual in the dataset. In this context, comparing the
slopes of different lines gives information on how much inter-individual weight differences
change over experimental time. Concomitantly, comparing different line intercepts informs
on individual growth rates: lines with different intercepts indicate different growth rates,
because two animals with the same initial weight reach different final weights within the
same period. Although this approach allows indirect corroboration of whether individuals
with different final size have different growth rates, it does not permit estimation (mean
value ± standard error) of those parameters in the equation that describe individual growth
over time.

Generalized linear mixed models (GLMM) are a statistical tool that can complement growth
analyses, because they allow modelling of the large variability in individual size observed
within a culture population. GLMM can be applied to non-normal data in which random
effects are present (Zuur et al., 2009). By incorporating components that modify the
structure of variance, mixed models yield more-reliable estimators of model coefficients. In
addition, through certain variance and correlation structures, mixed models may produce
new parameters that estimate size-at-age variability and the time elapsed for two size
measures to be statistically unrelated. GLMM procedures include the validation of models
by visual inspection of residuals Montgomery & Peck, (1992); Draper & Smith, (1998.),
thereby assuring that regression assumptions are adequately met.

Cephalopod growth has some remarkable characteristics: (i) growth rates are among the
highest in metazoans (the highest in invertebrate metazoans, higher than those of fish and
similar to those of mammals) Calow, (1987); (ii) it lacks an asymptotic growth phase
Moltschaniwskyj, (2004); (iii) it is highly plastic owing to its strong dependence on abiotic
and biotic factors, mainly temperature Pecl et al., (2004), the amount and quality of food
André et al., (2008), and sexual maturation Semmens et al., (2004); (iv) it follows a biphasic
pattern (as it often does in captivity), consisting of an initial rapid exponential phase
followed by a second phase, where growth slows down progressively (André et al., 2009);
and (v) it is highly variable intra-specifically Pecl et al., (2004); Leporati et al., (2007).

Size-at-age variability has been attributed to the lack of a strong association between age
and size of these soft bodied invertebrates, and is well documented both in wild and in
culture populations Leporati et al., (2007); Leporati et al., (2008). This great variability,
patent even under controlled temperature and food conditions, has been associated with

initial size Leporati et al., (2007), which in turn is highly variable both within and between broods. High coefficients of variation (>20%) in initial weight have been observed in several octopod species (*Octopus bimaculoides*, Forsythe & Van Heukelem, (1987); *Octopus diguetti*, DeRusha et al., (1989); *Octopus ocellatus*, Segawa & Nomoto, (2002). Intra-brood variation, however, could not be related to either the mother's weight or day of hatching Briceño-Jacques et al., (2010). Results suggest that growth rate does not depend on initial size in *O. maya* Briceño et al., (2010), but small differences in size during early life can be amplified and accumulated in time Vigliola & Meekan, (2009). This means that inter-individual variation in initial size has to be considered in experimental designs and data analyses aimed at understanding cephalopod growth.

Octopus maya is endemic to the Yucatan Peninsula Voss & Solis Ramirez, (1966), and its culture has received considerable attention Martinez et al., (2011); Uriarte et al., (2011). This species provides an interesting biological model to test hypotheses on heterogeneity in growth amongst cultured siblings, and serves to illustrate problems that arise in growth analysis and possible solutions to them.

Juveniles of *O. maya* from a single female were used. The eggs were held at 28°C (± 1°C) in an artificial incubator (without maternal care) until hatching. The weight of a total of 84 juveniles, that hatched over the course of 8 d, was recorded 24 h after hatching ($W_1 \pm 0.01$g), and then again on days 15, 45, 75 and 105. Octopuses were housed individually in 300 ml containers connected to a recirculation system in which water temperature was kept constant at 27 ± 1 °C. Octopuses were fed live adult brine shrimp (*Artemia salina*) and pieces of blue crab (*Callinectes sapidus*) meat *ad libitum*. Overall mortality (at day 105) was 60.7%.

To obtain a curve of weight gain as a function of time, a linear model was adjusted to the relationship between the natural logarithm of W_1, W_{15}, W_{45}, W_{75}, and W_{105} weights of all individuals and their corresponding ages:

$$Y_{ij} = \alpha + \beta_1 X_i + e_{ij}$$

Previous regression analysis, graphic representations of the data were explored to (i) identify extreme points (point graphs); (ii) assess normality (histograms and percentile graphs); and (iii) verify linear relationships X–Y graphs; Zuur et al., (2007). The regression was fitted with a generalized least square (GLS) procedure through restricted maximum likelihood (REML) and incorporated correlation and variance structures, using GLMM to ensure that homocedasticity and independence requisites were met (Fig. 9). Models featuring optimal correlation and variance structures were selected by considering values of the Akaike information criterion (AIC) and hypothesis tests based on F and likelihood ratio (L ratio) values. Once the significance of regression parameters were established through F, L ratio and t-tests, the model was validated by visual inspection of residuals Montgomery & Peck, (1992). We used the parameter δ as an estimator of the tendency of weight variances to increase with age.

Regression parameters differed significantly from 0 (Table 5). Interdependence of data over time (because we weighed animals repeatedly) resulted in a cyclic residual pattern, so we incorporated an autocorrelation structure (spherical spatial structure) in the random-effects term of the model (ε_i). Following Pinheiro & Bates, (2000), we kept this structure in the model, because (i) AIC values indicated that using it improved the model (AIC = 319.45),

Applied Ecophysiology: An Integrative Form to Know How Culture Environment Modulates
the Performance of Aquatic Species from an Energetic Point of View

205

and (ii) it made the cyclic residual pattern disappear. The estimated range parameter associated with this correlation structure was 74.3 (Table 5), representing the interval (in days) necessary to avoid correlation between two consecutive weight measurements of the same animal. Heterogeneity of variances was identified based on an increase in weight variation with age and was accounted for by including a variance structure of the type:

$$\varepsilon_i \sim N(0, \sigma^2 \bullet |age_i|^{2\delta})$$

Including this term significantly improved the model (L-ratio = 70.00; p < 0.001; AIC = 251.4). Finally, visual inspection of residuals revealed uniform dispersion and no evident patterns, and σ (a constant representing the increase in residual weight variation with age) was estimated to be 0.20 (Table 5).

Parameter	Value	Significance test
α	-2.182 ± 0.22 [a]	t = -100.92 [b]
β	0.030 ± 0.001	t = 48.60 [b]
σ	0.209	
δ	0.204	L.ratio = 70.00 [b]
Range	74.3	L.ratio = 114.20 [b]
AIC	251.4	

Table 5. Parameters of the exponential growth model: α = intercept; β = slope; σ = residual standard error; δ = variance structure parameter; range: correlation structure parameter defined by the variogram. Significance values of statistics used are also shown; [a] log-transformed values; [b] p < 0.001.

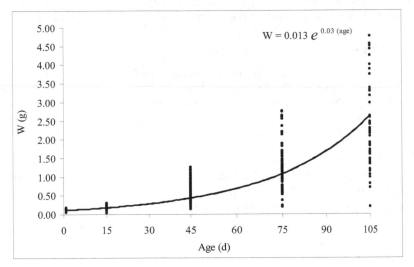

Fig. 9. Exponential relationship of weight and age (to age 105 d) in individually housed juvenile *Octopus maya* kept at 27 ± 1°C. Data points and continuous line are observed and predicted values, respectively.

The GLMM model included the use of a variance and a correlation structure that enhanced the model fit (the best AIC value was obtained: 251.4), and successfully explained octopus weight as a function of age. Because the date of hatching was registered for each individual, "age" is fixed in the model (i.e. it has no associated error), a feature in models fitted by least squares that is not usually met. According to this model, *O. maya* juveniles grow at a rate of 0.03 g/day, a rate similar to that obtained for other *O. maya* cultures (unpublished data). The δ parameter estimated in 0.204 is the exponent Given the correlation and variance structures included in the current model, and the parameters associated with them (δ, range), the procedures followed to adjust the GLMM model detailed here allow a high level of precision in predicting octopus weight from known age, thus, making the estimates of growth rate more reliable than other growth models.

6. Conclusions

During aquaculture technological development of species it is common that farmers or researchers start the culture only putting some animals in ponds, floating cages, or any other closed environment giving attention on productive parameters such as growth and survival. It is also common that, at the beginning of culture, animals are feed with the local available food in attempt to evaluate if in such conditions animals can be raised. However, this approach does not allow getting enough information for a sustainable development of culturing technology. In this sense, ecophysiology can help to evaluate how environmental conditions modulate several responses of cultured organisms, including the scope for growth, one of the key parameters involved in the energy balance equation. This model could be used as a base to make economical inferences or to modify culture methods. As showed in the present work, independently of the cultured species, the bulk of papers dedicated to demonstrate the effects of many environmental factors on physiological condition of animals validate applied ecophysiology as a useful method for the development and improvement of cultivation technology of aquatic organisms.

7. Acknowledgements

The present chapter is the result of many research made with the financial support of FONDEF D02i1163, D05i10217, Fondecyt 1110637, FONDEF D09I1153 in Chile and CONACYT through their financial support to make the sabbatical year of CR and Papiit IN 290327 to CR in Mexico.

8. References

Adachi, K., Hirata, T., Nishioka, T., Sakaguchi, M. (2003). Hemocyte components in crustaceans convert hemocyanin into a phenoloxidase-like enzyme. *Comparative Biochemical and Physiology*, 134B, 135-141.

Anderson, R.C., Wood, J.B. (2001). Enrichment for giagiant Pacific octopuses: happy as a clam? *Journal Applied Animal Welfare Science*, 4, 157-168.

André, J., Grist, E.P.M., Semmens, J.M., Pecl, G., Segawa, S. (2009). Effects of temperature on energetics and the growth pattern of benthic octopuses. *Journal of Experimental Marine Biology and Ecology*, 374, 167-179.

Applied Ecophysiology: An Integrative Form to Know How Culture Environment Modulates
the Performance of Aquatic Species from an Energetic Point of View

207

André, J., Pecl, G., Semmens, J.M., Grist, E.P.M. (2008). Early life-history processes in benthic octopus: Relationships between temperature, feeding, food conversion, and growth in juvenile *Octopus pallidus*. *Journal of Experimental Marine Biology and Ecology*, 354, 81-92.

Anger, K., Lovrich, G., Thatje, S., Calcagno, J. (2004). Larval and early juvenile development of *Lithodes santolla* (Molina, 1782) (Decapoda: Anomura: Lithodidae) reared at different temperatures in the laboratory. *Journal of Experimental Marine Biology and Ecology*, 306, 217-230.

Arena, L., Cuzon, G., Pascual, C., Gaxiola, G., Soyez, C., VanWormhoudt, A., Rosas, C. (2003). Physiological and genetic genetic variations in domesticated and wild populations of *Litopenaeus vannamei* fed with different carbohydrtaes levels. *Journal of Shellfish Research*, 22, 1-11.

Boletzky, S.V. (2003). Biology of early life in cephalopod molluscs. *Advances in Marine Biology*, 44, 144-203.

Boyd, C.E., (1995). Soil and water quality management in aquaculture ponds. INFOFISH International. Kuala Lumpur [INFOFISH INT.], 29-32.

Bradley, E.A. (1974). Some observations of *Octopus joubini* reared in an inland aquarium. *Journal of Zoology London*, 173, 355-368.

Bray, W.A., Lawrence, A.L., Leung-Trujillo, J.R. (1994). The effect of salinity on growth and survival of *Penaeus vannamei*, with observations on the interaction of IHHN virus and salinity. *Aquaculture*, 122, 133-146.

Briceño-Jacques, F., Mascaró, M., Rosas, C. (2010). GLMM-based modelling of growth in juvenile *Octopus maya* siblings: does growth depend on initial size? *ICES Journal of Marine Science*, 67, 1509-1516.

Brito, R., Chimal, Rosas, C. (2000). Effect of salinity in survival, growth and osmitic capacity of early juveniles of *Farfantepenaeus brasiliensis* (decapoda; penaeidae). *Journal of Experimental Marine Biology and Ecology* 244, 253-263.

Calow, P., (1987). Evolutionary physiological ecology? In: *Evolutionary physiological ecology*, Calow, P. (Ed.),. Cambridge University Press, New York, pp. 1-7.

Calcagno, J.A., Anger, K., Lovrich, G.A., Thatje, S., Kaffenberger, A. (2004). Larval development of the subantarctic king crabs *Lithodes santolla* and *Paralomis granulosa* reared in the laboratory. *Helgoland Marine Research*, 58, 11-14.

Charmantier, G., Soyez, C. and Aquacop (1994). Effect of moult stage and hypoxia on osmoregulatory capacity in the peneid shrimp *Penaeus vannamei*. *Journal of Experimental Marine Biology and Ecology* 178, 223–246

Chen, J.C. and Kou, T.T. (1998) Hemolymph acid–base balance, oxyhemocyanin, and protein levels of *Macrobrachium rosenbergii* at different concentrations of dissolved oxygen. *Journal of Crustacean Biology*, 18, 437–441

Cheng, W., Chen, J.C. (2000). Effects of pH, temperature and salinity on immune parameters of the freshwater prawn *Macrobrachium rosenbergii*. *Fish and Shellfish Immunology* 10, 387–391

Claybrook, D.L. (1983). Nitrogen metabolism. In: *The Biology of Crustacea* L.H, M. (Ed.), Internal anatomy and physiological regulation. Academic Press, New York, pp. 163-213.

Cortez, T., Gonzalez, A.F., Guerra, A. (1999). Growth of cultured *Octopus mimus* (Cephalopoda: Octopodidae). *Fisheries Research*, 40, 81-89.

Chen, J.C., Lin, M.N., Lin, J.Y., Ting, Y.Y. (1992). Effect of salinity on growth of *Penaeus chinensis* juveniles. *Comparative Biochemical and Physiology*, 102A, 343-346.

Danbom, D.B. (1979.) *The resisted revolution: urban America and the industrialization of agriculture, 1900 - 1930*. Iowa State University Press, Ames.

Danford, A., Uglow, R.F., Rosas, C., .Physiological responses of blue crabs (*Callinectes* sp) to procedures used in the soft crab fishery in La Laguna de T,rminos, México. In: B.C. Paust and A.A. Rice (eds). Marketing and Shipping Live Aquatic Products: Proceedings of the Second International Conference and Exhibition, November 1999, Seatle Washington, University of Alaska Sea Grant, AK-SG-01-03, Fairbanks: 1-8

Decamp, O., Cody, J., Conquest, L., Delanoy, G., Tacon, A. (2003). Effect of salinity on natural community and production of *Litopenaeus vannamei* (Boone), within experimental zero-water exchange culture systems. *Aquaculture Research*, 34, 345-355.

DeRusha, R.H., Forsythe, J.W., DiMarco, F.P., Hanlon, R.T. (1989). Alternative diets for maintaining and rearing cephalopods in captivity. *Laboratory Animal Science*, 39, 306-312.

Destoumieux, D., Moꚛoz, M., Bulet, P., Bachere, E. (2000). Peneidins, a family of anitimicrobial peptides from penaeid shrimp (Crustace, Decapoda). *Cellular and Molecular Life Sciences*, 57, 1260-1271.

Destoumieux, D., Saulnier, D., Garnier, J., Jouffrey, C., Bulet, P., Bachere, E. (2001). Antifungal peptides are generated from the C terminus of shrimp hemocyanin in response to microbial challenge. *The Journal of Biological Chemistry*, 276, 47070-47077.

Draper, N., Smith, H. (1998). *Applied Regression Analysis*. John Wiley, New York.

FAO (2011). *El estado mundial de la pesca y la acuicultura 2010*. Organización de las Naciones Unidas para la agricultura y alimentación. Departamento de Pesca y Acuicultura de la FAO.

Farías, A., García-Esquivel, Z., Viana, M.T. (2003). Physiological energetics of green abalone *Haliotis fulgens*, fed on a balanced diet. *Journal of Experimental Marine Biology and Ecology*, 289, 263-276.

Farias, A., Uriarte, I. (2006). Nutrition in pectinids. In: Scallops: *Biology, ecology and aquaculture*, Shumway, S.E., Parsons, G.J. (Eds.). Elsevier, pp. 521-542.

Farías, A., Uriarte, I. (2001). Effect of microalgae protein on the gonad development and physiological parameters for the scallop *Argopecten purpuratus* (Lamarck, 1819). *Journal of Shellfish Research*, 20, 97-105.

Farías, A., Uriarte, I., Castilla, J.C. (1998). A biochemical study of the larval and postlarval satges of the Chilean scallop *Argopecten purpuratus*. *Aquaculture*, 166, 37-47.

Farias, A., Uriarte, I., Hernández, J., Pino, S., Pascual, C., Caamal, C., Domingues, P., Rosas, C., 2009. How size relates to oxygen consumption, ammonia excretion, and ingestion rates in cold (*Enteroctopus megalocyathus*) and tropical (*Octopus maya*) octopus species. *Marine Biology*, 156, 1547-1558.

Applied Ecophysiology: An Integrative Form to Know How Culture Environment Modulates
the Performance of Aquatic Species from an Energetic Point of View

209

Farrell, A.P. (Ed). (2011). *Enciclopedia of fish physiology: from genome to environment.* Academic Press, New York.

Ferreiro, M.J., Pérez-Camacho, A., Labarta, U., Beiras, R., Planas, M.D., Férnandez-Reiriz, M.J. (1990). Changes in the biochemical composition of *Ostrea edulis* larvae fed on different food regimes. *Marine Biology,* 106, 395-401.

Flores, E.E.C. (1983). Visual discrimination testing in the squid *Tadores pacificus*: experimental evidence for lack of color vision. *Memoires of the National Museum Victoria,* 44, 213-227.

Forsythe, J.W., Van Heukelem, W.F. (1987). Growth. In: *Cephalopod life cycles,* Boyle, P.R. (Ed.),. Academic Press, London, pp. 135-155.

García-Garrido, S., Domingues, P., Navarro, J.C., Hachero-Cruzado, I., Garrido, D., Rosas, C. (2011). Growth, partial energy balance, mantle and digestive gland lipid composition of *Octopus vulgaris* (Cuvier, 1797) fed with two artificial diets. *Aquaculture Nutrition* 17, e174-e187.

Hall, M., Wang, R., Van Antwerpen, R., Sottrup-Jensen, L., S"derhall, K. (1999). The crayfish plasma clotting protein: a vitellogenin-related protein responsible for clot formation in crustacean blood. *Proceedings of the National Academy of Sciences USA,* 96, 1965-1970.

Hanlon, R.T., Turk, P.E., Lee, P.G. (1991). Squid and cuttlefish mariculture: An update perspective. *Journal of Cephalopod Biology,* 2, 31-40.

Hanlon, R.T., Wolterding, M.R. (1989). Behavior, body patterning, growth and life history of *Octopus briareus* cultured in the laboratory. *American Malacological Bulletin,* 7, 21-45.

Hernández, J., Uriarte, I., Viana, M.T., Westermeier, R., Farias, A. (2009). Growth performance of weaning red abalone (*Haliotis rufescens)* fed with *Macrocystis pyrifera* plantlets and *Porphyra columbina* compared with a formulated diet. *Aquaculture Research,* 40, 1694-1702.

Hochachka, P.W. and Somero, G.N. (2002) *Biochemical adaptation: mechanism and process in physiological evolution.* New York: Oxford University Press

Hochner, B., Shomrat, T., Fiorito, G. (2006). The octopus: a model for a comparative analysis of evolution of learning and memory mechanisms. *Biological Bulletin,* 210, 308-317.

Johnson, I. and Uglow, R.F. (1985) Some effects of aerial exposure respiratory physiology and blood chemistry of *Carcinus maenas* (L) and *Liocarcinus puber* (L). *Journal of Experimental Marine Biology and Ecology,* 94,1–12

Kattner, G., Graeve, M., Calcagno, J.A., Lovrich, G.A., Thatje, S., Anger, K. (2003). Lipid, fatty acid and protein utilization during lecithotrophic larval development of *Lithodes santolla* (Molina) and *Paralomis granulosa* (Jacquinot). *Journal of Experimental Marine Biology and Ecology,* 292, 61-74.

Iglesias, J., Sanchez, F.J., Bersano, J.F.G., Carrasco, J.F., Dhont, J., Fuentes, L., Linares, F., Muñoz, J.L., Okumura, K., Roo, J., van der Meeren, T., Vidal, E., Villanueva, R. (2007). Rearing of *Octopus vulgaris* paralarvae: Present status, bottlenecks and trends. *Aquaculture,* 266, 1-15.

Katsanevakis, S., Stephanopoulou, S., Miliou, H., Moraitou-Apostolopoulou, M., Verriopoulos, G. (2005). Oxygen consumption and ammonia excretion of *Octopus vulgaris* (Cephalopoda) in relation to body mass and temperature. *Marine Biology*, 146, 725-732.

Leporati, S., Pecl, G.T., Semmens, J.M. (2007). Cephalopod hatchling growth: the effects of initial size and seasonal temperatures. *Marine Biology*, 151, 1375-1383.

Leporati, S.C., Pecl, G.T., Semmens, J.M. (2008). Reproductive status of *Octopus pallidus*, and its relationship to age and size. *Marine Biology*, 155, 375-385.

Levin, L.A. (2003) Oxygen minimum zone benthos: adaptation and community response to hypoxia. *Oceanography and Marine Biology: An Annual Review*, 41,1- 45

Lovatelli, A., Farías, A., Uriarte, I. (Eds.) (2007). *Estado actual del cultivo y manejo de mosluscos bivalvos y su proyección futura: factores que afectan su sustentabilidad en América Latina*. Taller técnico de la FAO, 20-24 de agosto 2007, Puerto Montt, Chile. FAO, Roma, Italy.

Lovrich, G.A., Thatje, S., Calcagno, J.A., Anger, K., Kaffenberger, A. (2003). Changes in biomass and chemical composition during lecithotrophic larval development of the southern king crab, *Lithodes santolla* (Molina). *Journal of Experimental Marine Biology and Ecology*, 288, 65- 79.

Lucas, A. (1993). *Bioénergétique Des Animaux Aquatiques*. Masson, Paris.

Mair, J.M. (1980). Salinity and water type preferences of four species of postlarval shrimp (Penaeus) from Wets México. *Journal of Experimental Marine Biology and Ecology*, 45, 69-82.

Martinez, E., Aguilar, M., Trejo, L., Hernández, I., Díaz-Iglesia, E., Soto, L.A., Sanchez, A. and Rosas, C, (1998). Lethal low oxygen disolved oxygen concentrations for postlarvae and early juvenile *Penaeus setiferus* at different salinities and pH. *Journal of the World Aquaculture Society*, 29(2):221–229

Martinez, R., Lopez-Ripoll, E., Avila-Poveda, O., Santos-Ricalde, R., Mascaro, M., Rosas, C. (2011). Cytological ontogeny of the digestive gland in post-hatching *Octopus maya* , and cytological background of digestion in juveniles. *Aquatic Biology*, 11, 249-261.

Mayzaud, P. and Conover, R. (1988). O:N atomic ratio as a tool to describe zooplankton metabolism. *Marine Ecology Progress Series*, 45, 289-302.

McFarland, W.N., Lee, B.D. (1963). Osmotic and ionic concentrations of penaeid shrimps of Texas. *Bulletin of Marine Science of the Gulf and Caribbean* ,13, 391-417.

McGraw, W.J., Davies, D.A., Teichert-Coddington, D.R., Rouse, D.B. (2002). Acclimation of *Litopenaeus vannamei* postlarvae to low salinity: Influence of age, salinity endopoint and rate of salinity reduction. *Journal of the World Aquaculture Society* 33, 78-84.

Miliou, H., Fintikaki, M., Kountouris, T., Verriopoulos, G. (2005). Combined effects of temperature and body weight on growth and protein utilization of the common octopus *Octopus vulgaris*. *Aquaculture*, 249, 245-256.

Moltschaniwskyj, N.A. (2004). Understanding the process of growth in cephalopods. *Marine and Freshwater Research*, 55, 379–386.

Applied Ecophysiology: An Integrative Form to Know How Culture Environment Modulates
the Performance of Aquatic Species from an Energetic Point of View

211

Montaño-Pérez, K., Yepiz-Plascencia, G., Higuera-Ciapara, I., Vargas-Albores, F. (1999). Purification and characterization of the clotting protein from the white shrimp *Penaeus vannamei*. *Comparative Biochemical and Physiology*, 122, 381-387.

Montgomery, D.C., Peck, E.A. (1992). *Introduction to Linear Regression Analysis*. John Wiley, New York.

O'Dor, R.K., Wellls, M.J. (1987). Energy and nutrient flow. In: *Cephalopod life cycles*, O'Dor, R.K., Wellls, M.J. (Eds.),. Academic Press, London, pp. 109-131.

Ogle, J.T., Beaugez, K., Lotz, J.M. (1992). Effects of salinity on survival and growth of postlarval *Peaneus vannamei*. *Gulf Research Reports*, 8, 415-421.

Palacios, E., Bonilla, A., Perez, A., Racotta, I.S., Civera, R. (2004). Influence of highly unsaturated fatty acids on the responses of white shrimp (*Litopenaeus vannamei*) postlarvae to low salinity. *Aquaculture*, 299 201-215.

Paschke, K., Cumillaf, J.P., Loyola, S., Gebauer, P., Urbina, M., Chimal, M.E., Pascual, C. and Rosas, C. (2010). Effect of dissolved oxygen level on respiratory metabolism, nutritional physiology, and immune condition of southern king crab *Lithodes santolla* (Molina, 1782) (Decapoda, Lithodidae). *Marine Biology*, 157:7-18.

Pascual, C., Gaxiola, G., Rosas, C. (2003). Blood metabolites and hemocyanin of the white shrimp *Litopenaeus vannamei* : the effect of culture conditions and a comparison with other crustacean species. *Marine Biology*, 142, 735-745.

Pascual, C., Arena, L., Cuzon, G., Gaxiola, G., Taboada, G., Valenzuela, M., Rosas, C. (2004a). Effect of a size-based selection program on blood metabolites and immune response of *Litopenaeus vannamei* juveniles fed different dietary carbohydrate levels. *Aquaculture*, 230, 405-416.

Pascual, C., Zenteno, E., Cuzon, G., Sanchez, A., Gaxiola, G., Taboada, G., Suarez, J., Maldonado, T., Rosas, C. (2004b). Litopenaeus vannamei juveniles energetic balance and immunological response to dietary protein. *Aquaculture*, 236 431-450

Pascual, C., Sanchez, A., Zenteno, E., Cuzon, G., Gaxiola, G., Brito, R., Gelabert, R., Hidalgo, E., Rosas, C. (2006). Biochemical, physiological, and immunological changes during starvation in juveniles of *Litopenaeus vannamei*. *Aquaculture*, 251, 416-429.

Pauly, D. (1998). Why squid, though not fish, may be better understood by pretending they are. *South African Journal of Marine Science*, 20, 47-58.

Pecl, G., Steer, M.A., Hodgson, K. (2004). The role of hatchling size in generating the intrinsic size-at-age variability of cephalopods: extending the *Forsythe hypothesis*. *Marine Freshwater Research*, 55.

Perazzolo, L.M., Barracco, M.A. (1997). The prophenoloxidase activating system of the shrimp *Penaeus paulensis* and associated factors. *Developmental and Comparative Immunology* , 21, 385-395.

Pinheiro, J.C., Bates, D.M. (2000). *Mixed-Effects Models in S and S-Plus*. Springer, New York.

Ponce-Palafox, J., Mertjnez-Palacios, C.A., Ross, L.G. (1997). The effects of salinity and temperature on the growth and survival rates of juveniles white shrimp *Penaeus vannamei* , Boone, 1931. *Aquaculture*, 157, 107-115.

Pörtner, H.O., Storch, D., Heilmayer, O. (2005). Constraints and trade-offs in climate-dependent adaptation: energy budgets and growth in a latitudinal cline. *Scientia Marina*, 69 (Suppl. 2), 271-285

Rosas, C., Bolongaro-Crevenna, A., Sanchez, A., Gaxiola, G., Soto, L., Escobar, E. (1995). Role of the digestive gland in the energetic metabolism of *Penaeus setiferus*. *Biological Bulletin*, 189, 168-174.

Rosas, C., Martínez E., Gaxiola G., Brito R., Díaz-Iglesia E., Soto L.A. (1998). Effect of dissolved oxygen on the energy balance and survival of *Penaeus setiferus* juveniles. *Marine Ecology Progress Series*, 174, 67-75.

Rosas, C., Martinez, E., Gaxiola, G., Brito, R., Sanchez, A., Soto, L.A. (1999). The effect of dissolved oxygen and salinity on oxygen consumption, ammonia excretion and osmotic pressure of *Penaeus setiferus* (Linnaeus) juveniles. *Journal of Experimental Marine Biology and Ecology*, 234, 41-57.

Rosas, C., Cuzon, G., Gaxiola, G., Arena, L., Lemaire, P., Soyez, C., Van Wormhoudt, A. (2000). Influence of dietary carbohydrate on the metabolism of juvenile *Litopenaeus stylirostris*. *Journal of Experimental Marine Biology and Ecology*, 249, 181-198.

Rosas, C., Cuzon, G., Gaxiola, G., LePriol, Y., Pascual, C., Rossignyol, J., Contreras, F., Sanchez, A., Van Wormhoudt, A. (2001a). Metabolism and growth of juveniles of *Litopenaeus vannamei*: effect of salinity and dietary carbohydrate levels. *Journal of Experimental Marine Biology and Ecology*, 259 1-22.

Rosas, C., Cuzon, G., Taboada, G., Pascual, C., Gaxiola, G., Van Wormhoudt, A. (2001b). Effect of dietary protein and energy levels (P/E) on growth, oxygen consumption, hemolymph and digestive gland carbohydrates, nitrogen excretion and osmotic pressure of *Litopenaeus vannamei* and *L. setiferus* juveniles (Crustacea, Decapoda ; Penaeidae). *Aquaculture Research*, 32 1-20.

Rosas, C., Cuzon, G., Gaxiola, G., Pascual, C., Taboada, G., Arena, L., VanWormhoudt, A. (2002). An energetic and conceptual model of the physiological role of dietary carbohydrates and salinity on *Litopenaeus vannamei* juveniles. *Journal of Experimental Marine Biology and Ecology* 268, 47-67.

Rosas, C., Cuzon, G., Pascual, C., Gaxiola, G., Lopez, N., Maldonado, T., Domingues, P. (2007). Energy balance of *Octopus maya* fed crab and artificial diet. *Marine Biology*, 152, 371-378.

Scheel, D. (2002). Characteristics of habitats used by *Enteroctopus dofleini* in prince William Sound and Cook Inlet, Alaska. *Marine Ecology*, 23, 185-206.

Schmitt, A.S.C., Uglow, R.F. (1997). Hemolymph constituent levels and ammonia efflux rates of *Nephrops norvegicus* during emersion. *Marine Biology*, 127, 403-410.

Segawa, S., Nomoto, A. (2002). Laboratory growth, feeding, oxygen consumption and ammonia excretion of *Octopus ocellatus*. *Bulletin of Marine Science*, 71, 801-813.

Semmens, J.M., Pecl, G.T., Villanueva, R., Jouffre, D., Sobrino, I., Wood, J.B., Rigby, P.R. (2004). Understanding octopus growth: patterns, variability and physiology. *Marine and Freshwater Research* 55, 367–377.

Smith, C.D. (2003). Diet of *Octopus vulgaris* in False Bay, South Africa. *Marine Biology*, 143, 1127-1133.

Solis-Ramírez, M.J. (1997). The Octopus maya fishery of the Yucatán Peninsula. In: *The Fishery and Market Potential of Octopus in California*, Hochberg, L.,Ambrose, E. (Eds.), CMSC: pp. 1-10. Los Angeles, CA., pp. 1-10.

Applied Ecophysiology: An Integrative Form to Know How Culture Environment Modulates
the Performance of Aquatic Species from an Energetic Point of View

213

Sousa Reis, C., Fernándes, R. (2002). Growth observations on *Octopus vulgaris* Cuvier 1797 from the Portuguese waters: growth lines in the vestigial shell as possible tools for age determination. *Bulletin of Marine Science*, 71, 1099-1103.

Sritunyalucksana, K., Soderhall, K. (2000). The proPO and clotting system in crustaceans. *Aquaculture*, 191, 53-69.

Staples, D.J., Heales, D.S. (1991). Temperature and salinity optima for growth and survival of juveniles banana prawns *Penaeus merguiensis*. *Journal of Experimental Marine Biology and Ecology*, 154, 251-274.

Summer, W.C., McMahon, J.J. (1970). Survival of unfed squid, *Loligo pealei* in an aquarium. *Biological Bulletin*, 138, 389-396.

Sykes, A.V., Domingues, P., Correia, M., Andrade, P. (2006). Cuttlefish culture - state of art and future trends. *Vie et Millieu*, 56, 129-137.

Taylor, H.H. and Anstiss, J.M. (1999). Copper and haemocyanin dynamics in aquatic invertebrates. *Marine Freshwater Research*, 48,889–897

Thatje, S., Calcagno, J.A., Lovrich, G.A., Sartoris, F.J., Anger, K. (2003). Extended hatching periods in the subantarctic lithodid crabs *Lithodes santolla* and *Paralomis granulosa* (Crustacea: Decapoda: Lithodidae). *Helgoland Marine Research*, 57, 110-113.

Uriarte, I., Farías, A. (1999) The effect of dietary protein content of growth and biochemical composition of Chilean scallop *Argopecten purpuratus* (L.) postlarva and spat. *Aquaculture*, 180, 119-127.

Uriarte, I., Iglesias, J., Rosas, C., Viana, M.T., Navarro, J.C., Seixas, P., Vidal, E., Ausburger, A., Pereda, S., Godoy, F., Paschke, K., Farias, A., Olivares, A., Zuñiga, O. (2011). Current status and bottle neck of octopod aquaculture: the case of American species. *Journal of the World Aquaculture Society* In press.

Vargas-Albores, F., Yepiz-Plascencia, G. (2000). Beta glucan binding protein and its role in shrimp immune response. *Aquaculture* , 191, 13-21.

Vigliola, L., Meekan, M.G. (2009). The back-calculation of fish growth from otoliths. . In: *In Tropical Fish Otoliths: Information for Assessment, Management and Ecology*, Green, B.S., Mapstone, B.M., Carlos, G., Begg , G.A. (Eds.), Springer, New York, pp. 174–211.

Villanueva, R. (1993). Diet and mandibular growth of *Octopus magnificus* (Cephalopoda). *South African Journal of Marine Science* 13, 121-126.

Villanueva, R., Norman, M.D. (2008). Biology of the planktonic stages of benthic octopuses. *Oceanography and Marine Biology - Annual Review* 46, 105-202.

Voss, G.L., Solis Ramirez, M.J., 1966. *Octapus maya*, a new species from the Bay of Campeche, Mexico. *Bulletin Marine Science*, 16, 615-625.

Weiner, D.R. (2000). *Models of Nature: Ecology, Conservation and Cultural Revolution in Soviet Russia*. University of Pittsburg Press, USA.

Wieser,W. (1986). *Bioenergetik. Energietransformationen bei Organismen*. Georg-Thieme-Verlag, Stuttgart, 245 pp.

Wollesen, T., Loesel, R., Wanninger, A. (2009). Pygmy squids and giant brains: Maping the compley cephalopod CNS by phalloidin staining of vibratome sections and whole-mount preparations. *Journal of Neuroscience Methods*, 179, 63-67.

Zar, J.H. (1999). *Biostatistical Analysis*. Prentice-Hall, New Jersey.

Zuur, A.F., Ieno, E.N., Smith, G.M. (2007). *Analysing Ecological Data Series: Statistics for Biology and Health.* Springer, New York.

Zuur, A.F., Ieno, E.N., Walker, N.J., Saveliev, A.A., Smith, G.M. (2009). *Mixed Effects Models and Extensions in Ecology with R.* Springer, New York.

11

Chilean Salmon Farming on the Horizon of Sustainability: Review of the Development of a Highly Intensive Production, the ISA Crisis and Implemented Actions to Reconstruct a More Sustainable Aquaculture Industry

Pablo Ibieta[1], Valentina Tapia[1], Claudia Venegas[1],
Mary Hausdorf[1] and Harald Takle[1,2]
[1]AVS Chile SA, Puerto Varas
[2]Nofima, Ås
[1]Chile
[2]Norway

1. Introduction

Historically the Chilean economy has been based in exports accounting for 35% of the gross domestic product and mining is the main export income in the country. However, since 1980s as a part of a promotion the Chilean economy has diversified in part away from its dangerous over-reliance on copper exports. The growth of earnings in the fruit, wine, wood and forestry, fisheries and aquaculture sectors in particular, has been rapid since the early 1980s as Chile has exploited its comparative advantage in environmental endowment and low labour costs on the global market (Barton, 1997; Barton & Murray, 2009). The salmonid cultivation is restricted to regions with particular water temperature ranges in both fresh and seawater environments, sheltered waters and critically excellent water quality. Thus, salmonid aquaculture has become an important activity and the main development gear in the southern regions of Chile. The Chilean salmon industry has shown a fast development over the last 20 years and, therefore; today, Chile is the largest producer of farmed rainbow trout and Coho salmon, and the second worldwide of Atlantic salmon. This situation is the result from the application of innovation and development of value added products, which produced an average annual growth rate of 22% from 1990 to 2007 and an increase of exports from USD 159 million in 1991 to USD 2,242 million in 2007 (SalmonChile, 2007a). In 2006, salmon exports represented 12.94% of the non-mining-related national exports and 38.75% of the Chilean food exports that decreased due to a sanitary crisis down to 9.94% and 31.08% in 2010, respectively (Banco Central de Chile, 2010). The latter indicated that the salmon industry has become an important factor in economic diversification and one fundamental base in the strategy towards positioning Chile in the top rank of world food producers (SalmonChile, 2007b). The growth of aquaculture also impact positively other important sectors of the national economy, such as transport, processing, feed and engineer suppliers, laboratories, veterinary services and many others.

The success of the salmon aquaculture in Chile has been the product of the appropriate assimilation of foreign technologies and development of local technological capabilities. Although national investors played a major role in the early development phases of the industry, the entry of large foreign companies in the last two decades has facilitated the introduction of technologies, enlargement of production, vertical integration, merging and increasing the size of companies. This industry has also contributed to the general development of the economically depressed and rural regions in southern Chile. Although the Chilean salmon aquaculture has performed an astonishing development over the last 20 years, there have been demonstrated severe knowledge gaps before and during the devastating sanitary crisis caused by the infectious salmon anaemia (ISA) outbreaks that nearly led to a collapse of the industry.

The aim of the present chapter is to give an overview of the salmon aquaculture in Chile. Firstly, we describe how salmon farming rose up in a country without native salmonids, and the succeeding establishment of a highly organized and globalized industry. Secondly, we review the congestion of disease problems that peaked with the ISA outbreak that nearly collapsed the whole Chilean salmon industry, and finally the strategies and concrete measurements that have been implemented by the authorities and the industry to remerge the salmon aquaculture in Chile.

2. The Chilean salmon industry

The interest in introducing salmon to the water bodies of Chile began in second half of the 19th century with the first import of salmon and trout eggs in 1885 (Bluth et al., 2003). In 1905, the first Atlantic salmon (*Salmo salar*) and rainbow trout (*Oncorhynchus mykiss*) eggs were imported and successfully produced in Chile (Bluth et al., 2003). Since 1920s government institutions carried out several attempts to introduce Pacific and Chinook salmon in lakes and rivers driven by the interest on sport fishing. However, it was not until 1969 when the program to introduce Pacific salmon in Chile was formalized by an agreement between the Japan International Cooperation Agency (JICA), the Fisheries Association of Japan, and the Chilean National Fishing Agency. As a part of the project, salmon fish farms with egg incubation facilities were constructed and also a training program carried out in Japan. In order to strengthen this sector, the Chilean government created the Office of the Undersecretariat for Fisheries (Subpesca) and the National Service for Fisheries (Sernapesca) under the Ministry of Economy, Development and Reconstruction in 1978. In 1974, the first private initiative to farm rainbow trout took place in Chile; hence, the company Sociedad de Pesquerías Piscicultura Lago Llanquihue Ltda had successfully exported trout to France, and soon after, Chilean farmed trout reached North America and other countries in Europe. During the 1980s, Fundación Chile, a private and non-profit organization played an essential role in the development of salmon aquaculture supporting technical and commercial issues toward large-scale salmon farming. It also focused on research and the implementation of new technology for raising salmon, artificial reproduction, behavioural studies and breeding, as well as the creation and exploitation of new fresh and seawater farm sites (Bluth et al., 2003). One of the final state initiatives to introduce salmon in Chile was through the cooperation between the Subpesca and the Canadian International Development Agency (CIDA). This was implemented by Hatfield Consultants Incorporate that technically advised many companies and fish farms and supported to consolidate companies operating as a technological bridge and providing services that helped the development of the salmon industry (Bluth et al., 2003).

2.1 Emergence of industrial aquaculture

The emergence of foreign companies in the salmon-farming business stimulated interest among local investors and firms in the commercial viability of the industry. Some of the first enterprises were founded by biologists, veterinarians and marine biologist experts. They had acquired substantial experience and knowhow regarding fish farming management of foreign species and production of fish eggs (Norambuena & González, 2006). In the late 1980s, Chile officially entered the group of salmon and trout producing countries and a number of local salmon farming companies increased and production grew tremendously. In 1985, 36 salmon farms were operating in Chile and the production reached near 1,200 tonnes increasing up to 60,000 tonnes by 1991 (Bjørndal, 2002). Thus, the industry grew in technology focused to farming, feed and fish processing. In the early stages, the production of salmonid in fresh and seawater has been centred in Los Lagos region. However, in the last 20 years, the development of connectivity, fish handling and transport technologies lead to a wide spread in egg, fry, parr and smolt production up north to Valparaíso region (V), and also fresh and seawater production further south to Magallanes region (XII), see Figure 1.

2.2 Structure and organization of the industry

The salmon industry emerged as a mature cluster, with several companies undertaking different aspects of salmon production and marketing. In Chile there has been a strong tendency towards vertical integration in the production of salmonids from egg production to market. Even minor producers will process, market and export their own production. Furthermore, it is common for farmers to have two or three salmon species, in order to spread the risk, both on more species and on more markets. This also contributes to smoother harvesting patterns and consequently cash flow, throughout the year. Large companies have moved to vertical integration in order to reduce production costs, implementing egg production, feeds and processing plants within their business (Iizuka, 2004; Norambuena & González, 2006).

Interestingly some farming companies are now producing most of their feed in-house or are part of holding-groups, thus part of the feed is addressed to the fish farmers within the holding. The Chilean industry has organized merges differently from that seen in the European salmon industry. After merging process towards larger enterprises, the original smaller companies were not integrated completely. Hence, they have changed the ownership and some restructuring, but kept most of the structure in each daughter companies and the original names.

The number of companies controlling most of the salmonid production in Chile has decreased from 35 to 10 companies between 1997 and 2006. However, today 19 companies contribute with more than 80% of the production of salmonids (Table 2).

More than 60% of the employment in the salmon industry is within processing and value added production today. About 90% of the Atlantic salmon production and 30% of Coho salmon and rainbow trout production is processed in the country. The main centres for processing salmonids are in the southern regions around Puerto Montt and Quellón area.

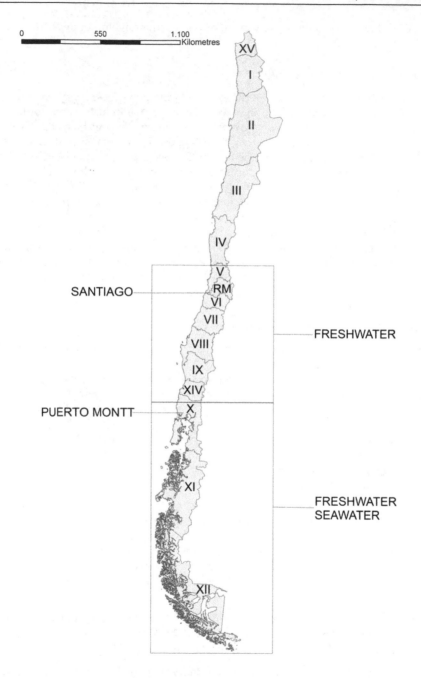

Fig. 1. Map of Chile and areas related to salmon aquaculture activities (exclusively fresh water production and fresh/sea water production, MR: Metropolitan region) (map source: GeoAustralis).

The consolidation of the salmon farming industry was the creation of the Association of Salmon and Trout Producers of Chile (APSTC) gathering 17 companies in 1986. In 2002, the name changed to the Association of the salmon industry of Chile, SalmonChile A.G., which also includes supplier firms. Nowadays, SalmonChile A.G. has a total of 76 companies as partners, where 25 are solely salmonid producers. Furthermore, companies related to salmon production established their own associations. For instance, the Association of Net and Net Service Industries (ATARED), Regional Association of Ship-owners and Maritime Services (ARASEMAR) and Association of Diving Companies were all established in 2001 and represent more than 70% of each sector. In order to further strengthen and enhance competitiveness, in 1995 the Association of Salmon and Trout Producers created Intesal, the Salmon and Trout Technology Institute, with financial support from the Production Development Corporation of Chile, CORFO. This institution was established to develop, share and prospect quality control technologies and improved food safety in the salmon industry. Intesal provides technical assistance and training in sanitary and quality control standards for its partners. Currently, Intesal has an important role in supporting the institutional consolidation of the salmon cluster and the introduction of quality control management systems.

In 2009, Trout and Coho salmon producers created their own association (ACOTRUCH) in order to represent medium and small companies not producing Atlantic salmon and not directly affected by sanitary events triggered by ISAv.

The salmon cluster has an important role in the development of the industry, due to the increasing service requirements every year. Continuous investment has been necessary while industry moves forward to informatics and automation in processes. The National Statistic Institute (INE) estimated that around 1,200 companies take part in the salmon industry where of 500 are key actors, 100 as input suppliers and 400 as service companies.

Association	Members	N° of companies	Capacities
ARASEMAR	The Association of Regional Ship-Owners for Shipping Services	14	200 vessels and 16 well boats
ADEB	Association of Diving Companies	8	500 professional and 3000 not professional divers
ATARED	Association of Net Workshops	14 (30)*	Net manufacturing and maintenance
ALAVET	Association of Veterinary Laboratories	18	pharmaceutical labs**
ARMASUR	Association of Southern Ship-Owners	10	Ship builders
SALMONCHILE	Association of the Chilean Salmon Industry	76 (25***)	Farming companies and suppliers
ACOTRUCH	Association of Coho salmon and trout producers	10	Farming companies

* not associated; ** related to the salmon cluster; ***only producers/farmers

Table 1. The key members of the salmon cluster.

2.3 Salmon production in Chile

The larger fresh water (FW) production is based in La Araucanía region (IX), Los Ríos (XIV) andLos Lagos (X). Seawater (SW) production has been mainly developed in Los Lagos region; in 2006, on a rather small area from Puerto Montt in the north to Quellón on the Chiloé Island a total of 499,512 tons of salmonids were produced. Due to the increase of SW production in Los Lagos region, Quellón was expected to become the biggest centre for salmon processing in Chile. However, before the ISA crisis in 2008, the production has been expanded to Aysén region (XI) and Magallanes region (XII) having the total harvest of 208,961 tons and 6,053 tons, respectively. In 2005, SalmonChile A.G. projected an annual growth of about 8-10% for the following 5-10 years. These expectations were based on exploiting the potential of moving most of the production to Aysén and Magallanes regions and to expand the FW production using highly advanced recirculation plants.

Historically the salmon cluster, including producers, processing plants, and services, has demanded up to 28,368 and 7,631 direct and indirect job positions, respectively (SalmonChile, 2007b). Specifically, the industry covers 0.7% of the job positions in the country and accounted for 11% of the employment in the Los Lagos region (X). The latter has contributed significantly and positively to the improvement of poverty in those municipalities where salmon aquaculture has been developed. Usually, the employment within producers is distributed in each of these stages of production with approximately 65% of total workers in sea farms, 30% in smolt production and 5% in hatcheries producing eggs. However, the number of employees decreased drastically down to 50% due to the sanitary crisis and the lower production in 2009.

2.3.1 Freshwater production

Current FW operations in Chile include lake-based, tank and cage systems, estuary cage systems, stream-based flow-through systems and recirculation tank systems. In 2008, the Chilean FW farms produced over 741 million salmonid ova and 298 million smolts, out of which 29% came from estuary-based farms, 30% from lakes and 41% from land-based facilities, or nurseries. While most hatcheries supply smolts for grow-out in farms owned by their parent company, many hatcheries also contract to supply smolts to unrelated farms, either using eggs from their own broodstock lines or using eggs provided by the contracting farms (Olson & Criddle, 2008). There are 169 FW facilities in Chile out of which 20 are recirculation systems (Silva, 2010a). Historically, in Chile, to produce one smolt has required from 2 to 5 eggs doubling the current needs in Norway (Águila & Silva, 2008).

Chilean-spawned eggs were available as early as 1980, but Chilean hatcheries remained largely dependent on fertilized eggs from foreign broodstock through the 1990s. Since the year 2000, nevertheless, Chilean egg production increased dramatically (Figure 2) and 79% of all eggs used in Chilean hatcheries were produced from local broodstock in 2009 (Sernapesca, 2011a). The decrease of near 70% observed in the national production of Atlantic salmon eggs is based on the effects of the ISAv crisis, going from near 500 million eggs in 2008 to close to 160 million in 2009. In addition, the import of egg fell substantially in 2009, mainly due to the lower production, but also based on restrictions, fear of importing new pathogens and changes in production strategies implemented by different companies. However, the import of rainbow trout egg increased more than double in 2010. Both rainbow trout and Coho salmon production grew approximately 25% and 37% with an increase in production due to the lower sanitary risk involved in their production between 2009 and 2010.

Chilean Salmon Farming on the Horizon of Sustainability: Review of the Development of a Highly Intensive Production, the ISA Crisis and Implemented Actions to Reconstruct a More Sustainable Aquaculture Industry

221

	Total Exports (ton net)		Value (thousand USD FOB)		Price (USD/kg FOB)	
	2009	2010	2009	2010	2009	2010
Total	369,216.1	297,160.3	2,101,643.7	2,060,909.0	5.7	6.9
Aquachile SA (Aquachile, Salmones Chiloé, Salmones Maullín, Aguas Claras)	47,715.6	34,172.7	254,794.0	206,918.70	5.3	6.1
Mainstream Chile SA	33,927.1	23,607.4	172,820.8	136,795.60	5.1	5.8
Pesquera Los Fiordos	27,304.6	16,368.7	129,961.2	106,239.10	4.8	6.5
Marine Harvest Chile, Deli Fish SA	23,731.9	12,702.2	144,049.5	116,118.20	6.1	9.1
Multiexport Foods, Salmones Multiexport SA	23,358.8	21,046.7	120,780.6	164,539.10	5.2	7.8
Cia Pesquera Camanchaca SA	15,667.5	9,906.3	115,634.8	66,684.90	6.7	6.7
Salmones Antártica Ltda	14,600.0	13,847.9	98,780.7	102,868.20	6.8	7.4
Granja Marina Tornagaleones Ltda	12,478.1	11,911.6	63,182.4	62,948.90	5.1	5.3
Trusal SA	12,165.8	13,465.2	70,212.7	103,529.80	5.8	7.7
Salmones Cupquelan SA	12,083.5	8,842.7	70,252.7	72,912.80	5.8	8.2
Ventisqueros SA	10,613.5	12,550.1	70,247.6	90,768.40	6.6	7.2
Acuinova Chile SA	8,737.0	9,441.4	51,403.4	67,303.00	5.9	7.1
Salmones Pacific Star	8,614.8	8,845.6	42,578.1	49,275.50	4.9	5.6
Cultivos Marinos Chiloé SA	8,553.8	8,242.0	67,692.3	74,819.70	7.9	9.1
Salmones de Chile SA	6,318.4	7,438.0	33,671.9	43,918.10	5.3	5.9
Australis Mar SA	5,177.9	7,942.6	32,418.2	56,505.30	6.3	7.1
Caleta Bay Export Ltda	4,958.9	5,733.8	28,792.8	49,116.70	5.8	8.6
Salmones Itata SA	4,413.1	9,518.1	33,901.1	75,612.50	7.7	7.9
Salmones Aysén	3,345.9	9,827.7	15,926.3	54,930.60	4.8	5.6
Others	85,449.3	51,749.6	484,542.1	359,103.90	5.7	6.9

*source TechnoPress, 2010

Table 2. Ranking of salmonid exporting companies 2009-2010.

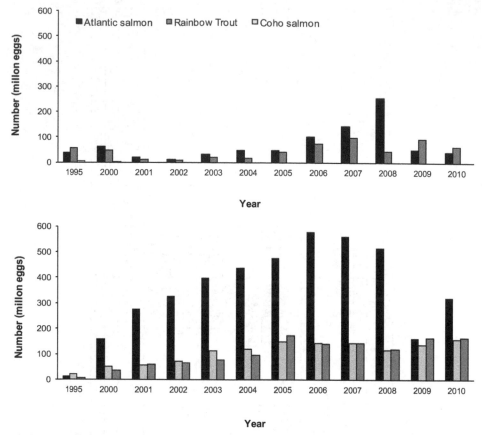

Fig. 2. Incubated imported eggs (upper) and incubated national eggs (bottom). Source: Sernapesca, 2011a

Various strategies have been used for smolt production in Chile. It was estimated that the majority of current smolt production comes primarily from flow-through fish farms that produce fry, followed by the smolting process in lake, river and estuary-based farms. No further growth in lake production is foreseen, since the regulation prohibits the authorization of new farms, considering that "Suitable Areas for Aquaculture" have not been designated in any Chilean lakes since year 1990. This means that the current production in lakes takes place on sites that were authorized prior to the aforementioned regulation, and that the growth of these farms is based solely on production increases in restricted spaces. Thus, a single company may have only one or a mix of smolt production strategies that range from a complete land based production to a mixture of using land based facilities, lake and estuarine areas (Figure 3).

Due to the existence of imported eggs, and offseason spawning, smolts are produced throughout the year (Figure 4) and smoltification is carried out using natural photoperiod, artificial lights and specialized feed additives. In contrast with the smolt types in Northern hemisphere; 1-year smolt (S1, S1 ½), 2-years smolt (S2) and half-years smolt (S ½ or 0+), in

Chilean Salmon Farming on the Horizon of Sustainability: Review of the Development of a Highly Intensive
Production, the ISA Crisis and Implemented Actions to Reconstruct a More Sustainable Aquaculture Industry

223

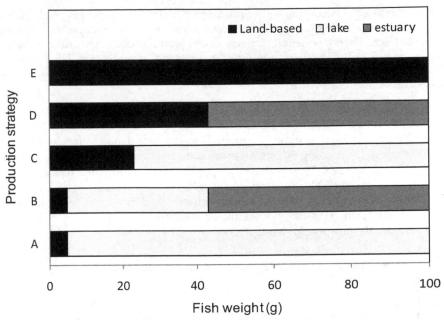

Fig. 3. The graphic shows 5 smolt production strategies in Chile. A. Eyed eggs to fry on land-based, fry to smolt on lake. B. Eyed eggs to fry on land-based, fry to fry on lake and fry to smolt on estuary, C. Eyed eggs to fry on land based and fry to smolt on lake, D. Eyed eggs to fry on land-based and fry to smolt on estuary, E. Eyed eggs to smolt on land-based facilities.

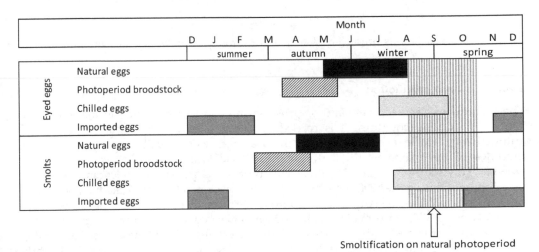

Fig. 4. Season and off-season smolt production year around in Chile. In the southern parts of Chile, the four year seasons are well defined with markedly differences between summer, autumn, winter and spring.

Chile smolts are called according to the production method as in-season smolt, late smolt, and early smolt, depending mainly on broodstock´s photoperiod manipulation, chilled egg, and imported egg. Therefore, smolt size regardless production method will normally range between 90 g and 120 g. Furthermore, the transportation of smolt over 250 g is prohibited and a certain small amount of smolts between 60 g and 90 g are produced by land-based facilities.

Smolt production reached its historical peak production in 2007, presenting a significant decrease in 2008 and 2009 (Figure 5), going from close to 400 million in 2007 to 160 million in 2009. This abrupt decrease in smolt production, mainly in Atlantic salmon, was caused by the sanitary crisis. Commonly smolts are transported by road and wellboat to their final ongrowing site in the sea. Due to the widespread FW areas in Chile fish are frequently transported on land over 300 kilometres prior wellboat transportation. In addition, fish may be transported to a lake or estuary then following land transport, wellboat and final ongrowing destination. This has been a major challenge in welfare and adequate water quality for the smolt transport systems.

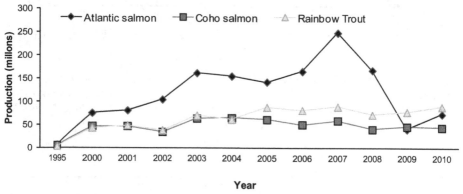

Fig. 5. Number of smolts produced (data available from year 1995). Source: Sernapesca 2011a

2.3.2 Seawater production

The technology used for salmonid on-growing in Chile is very similar to the rest of the salmon producing countries, it is carried out in cages and today most operations are automatized. At the end of the 1980s the sea cages were usually few square wooden modules of approximately 1.100 m³ (10x10x11) and located near-shore or in sheltered waters. However, the size, location and type of cages have undergone profound development during the 1990s and 2000s. Sea sites developed to over 12 modules of circular high-density polyethylene cages of 12.725 m³ (d. 30m) or square modules hinged steel cages of 25.600 m³ (40x40x16) provided with modern automated systems. Thus, large integrated sites are also provided with pontoons including all comfortable living needs. The feeding and fish-waste cleaning procedures have been changing enormously from 1980s. In the early days of the salmon aquaculture, mainly hand control was performed that today is carried out by automatic feeding systems, satellite and under water video devices. An important development in anti-predators devices have been carried out due to huge losses caused by

Chilean Salmon Farming on the Horizon of Sustainability: Review of the Development of a Highly Intensive
Production, the ISA Crisis and Implemented Actions to Reconstruct a More Sustainable Aquaculture Industry

225

sea lions. Nearly 90% of the Chilean salmon farms had reported attacks by sea lions and losses around USD140 million in 2009. Sea lion manages to cleave regular nets and avoid acoustic harassment, alarm devices and fake killer whales. Nowadays predator nets are the only effective method towards sea lion attacks (Oliva et al. 2003; Vilata et al. 2010).

The SW production have been spreading further south in Aysén region, where many concessions have been granted in the past years and activities are starting far south in Magallanes region. Today from 1063 concessions granted, 441 are in Los Lagos region, 572 in Aysén region and 50 in Magallanes region, however, a nearly 15 times growth is expected by 2013. In the Southern hemisphere there are inverse seasons in respect to Northern hemisphere, thus allowing the supply of fresh salmon during winter time in importing countries as USA and Japan. During summer, the temperature in the sea ranges between 8°C and 18°C in Los Lagos region, but in Aysén and Magallanes seldom reach up to 15°C. In addition, in winter temperatures could reach down to 5°C but never around freezing point.

2.3.3 Salmonids processing

Chilean salmon farming processing technology has evolved from fully-manual systems to a mixed semi-automated systems, where some phases of the system are automated and other phases are carried out manually to ensure quality standards. Nevertheless, processing facilities in Chile employ twice of the workers as equivalent facilities in competitor countries, due to the labour costs. Thus, providing a comparative advantage in labour-intensive activities such as the production of fillets and boneless portions of salmon. The main salmon products are fresh and frozen fillets, loins and portions, but also smoked, and marinated products are significant parts of production. There has been an increase in value-added products obtained from salmon, mainly in Atlantic salmon and rainbow trout. This triggered by intense competition in the generic salmon market and prices trend in value-added products. Each farming company has 2 to 4 major customers that define the product range; however, very often these customers bring special equipment and procedures for production of innovative products with the brand of these customers. The reaction to this situation has been differentiation: creation of brands, greater variety in processing (fresh, smoked, frozen, salted, canned salmon) production of related products (fresh or frozen fish portions of specific weight and sizes to fill the specific requirements of clients, salmon cuts designed to facilitate fish preparation and decrease waste, etc.), client labelling and packaging for final consumption, and food traceability. These actions however, have increased production costs (Perlman & Juárez-Rubio, 2010).

2.4 Chilean salmon market

In the late 1990s, the development of the industry was driven by market needs. The fall in salmon prices on the international market in the late 1980s and early 1990s led to collapse of the smaller companies and established a consolidated industrial aquaculture in salmon producing countries such as Chile, Norway and Scotland. Consequently, the Chilean salmon industry realised the risk of relying almost exclusively on two major markets; hence, a group of 13 local companies formed Salmocorp in order to face existing markets and explore new ones. Although this joint venture company operated only for three years, it certainly contributed to opening up new markets. Despite the main markets of Chilean salmon have traditionally been USA and Japan; in the last decade, new markets have been expanded to

Latin America as well as to Asia (Iizuka, 2004). Therefore, exports to these markets are expected to grow relatively faster in the following years than the established markets in Japan and USA (Table 3).

Year	1999	2005	2006	2007	2008	2009	2010
	USD (million)						
Japan	471	638	704	648	713	824	921
USA	259	606	792	862	795	554	466
EU	34	236	308	279	284	162	73
Latin America	39	88	156	202	268	290	357
Other (Asia)	15	153	246	250	333	270	314

Table 3. Main market shares of Chilean salmon aquaculture from year 1999 to 2010 (USD million). Source: TechnoPress, 2010.

2.5 Production cost of Chilean salmon aquaculture

The production cost of salmonids in Chile has been low in relation to other salmonid-producing countries (Figure 6). Lower costs for labour force and lower prices for feed ingredients have been the major factors for the cost advantage in Chile. Historically, Chilean feed producers could pay significant lower prices for fishmeal and fish oil in South America than producers in the Northern Hemisphere. However, with increasing fishmeal and fishoil prices, improving life standards and education the Chilean cost advantages may be reduced significantly in the future. Furthermore, the production cost will considerably increase because the new enforcements of the regulation from the year 2011.

3. Diseases in salmon farming in Chile

Unfortunately, in Chile the advantage of a disease-free environment for farmed fish has been gradually lost the past two decades. This situation is acknowledged and majorly explained by the introduction of exotic pathogens, mainly through imported eggs (Smith et al., 2001; Claude et al., 2000). Chile has imported more than 1,900 million eggs over the years from various countries and continents of the Northern hemisphere, Figure 2. It is known that when moving fish or their gametes, their pathogens are moved along with them. Notwithstanding, in the case of some diseases in salmon farming in Chile, a relatively opposite phenomenon could have been possible, namely that some unknown etiologic agents could be endemic in native species without causing apparent damage and have adapted to it. On the other hand, non-native fish species such as salmonids would not have a proper genetic makeup or effective defence mechanisms against these new organisms, presenting a diminished health status with consequent outbreaks. Intensive systems used in aquaculture i.e. huge biomass grown along with high densities of fish per unit of water volume, have contributed to increase the prevalence of pathogens and to increase contact rates with hosts, all of which translates into a higher risk of disease.

3.1 Time-course of disease emergence

The appearance of important diseases posing a risk to the salmon industry in Chile has been related directly to the increase in production (Figure 7), where volumes went from

Chilean Salmon Farming on the Horizon of Sustainability: Review of the Development of a Highly Intensive
Production, the ISA Crisis and Implemented Actions to Reconstruct a More Sustainable Aquaculture Industry

227

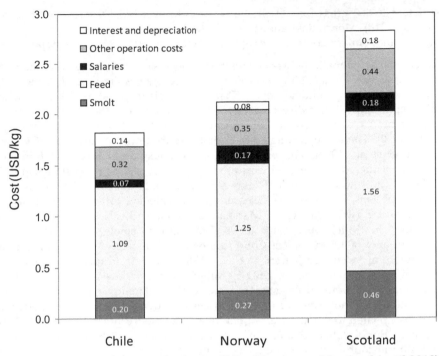

Fig. 6. Cost structure in salmon production in Chile, Norway and Scotland (until 2010)

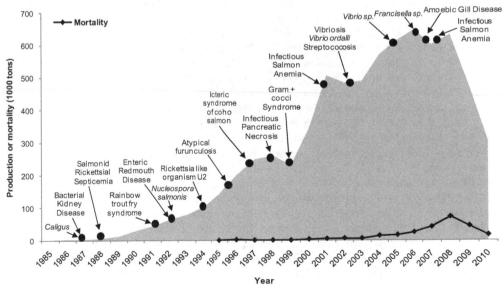

Fig. 7. Time line of production, disease emergence and mortality in salmon farming in Chile
(period 1985-2010). Source: Sernapesca, 2011b; TechnoPress, 2010.

approximately 10,000 gross tons in the early 1990s to over 545,000 tons exported in 2008 (TechnoPress, 2010). Mortalities caused by infectious diseases increased over the years and dropped after the sanitary crisis as less salmon where transferred for grow-out purposes.

In the early 1990´s, the main diseases present in Chilean salmon farming were the Bacterial Kidney Disease (BKD), affecting rainbow trout, Atlantic salmon and Coho salmon, throughout their entire production cycle. Piscirickettsiosis (Fryer et al., 1992) or Salmonid Rickettsial Septicemia (SRS), a disease detected for the first time in the world, endangered at that time mainly Coho salmon during grow-out phase (Bravo & Campos, 1989a, 1989b; Fryer et al., 1990). The etiologcal agent was first described in southern Chile (Cvitanich et al., 1991; Fryer et al., 1990). The pathogen was extremely aggressive, and farms in the southern part of the country were devastated over a very short period (Branson & Diaz-Muñoz, 1991). Until now, piscirickettsiosis caused by the facultative intracellular bacteria, *Piscirickettsia salmonis*, is one of the diseases most threatening to the sustainability of the Chilean salmon industry (Olivares & Marshall, 2009). Piscirickettsiosis affects all salmonid species farmed in Chile, causing up to 90% mortality in some sites. First calculations estimated mortality of 1,5 million of Coho salmon, which represented an estimated loss of USD 10 million for year 1989 (Larenas et al., 2000). According to further calculations, the industry attributes annual losses of USD 50 million to piscirickettsiosis (Smith et al., 1997). Currently data from SalmonChile estimate losses of USD 19 million considering direct costs due to mortalities but not considering indirect costs i.e. treatment, decreased performance, and other costs related to the disease. Global losses have been estimated to USD 120 million for each year (InnovaChile, 2008). Different factors contribute to the occurrence of a persistent clinical course, much more as a chronic presentation of the disease with abscess observation (Godoy, 2010). Today, piscirickettsiosis represents 90% of the infectious-disease-caused mortalities (Intesal pers. comm., 2011) and is the pathogen agent mostly diagnosed in SW (Subpesca, 2010). In addition to the appearance of BKD and SRS, the first intensive productions of eggs and fries were visibly affected by diseases caused by funguses (*Saprolegnia* sp.) and flavobacteria (mainly *Flavobacterium columnare* and *F. psycrophilum*) in hatcheries and fry farming sites located in lakes. These infections are considered the major problems in FW, where infections with flavobacteria can result in 5% to 70% mortality rate of fingerlings (Valdebenito & Avendaño-Herrera, 2009). In 2008, *Saprolegnia* showed a monthly average mortality in Coho salmon of 0.06%, in Atlantic salmon of 0.02% and in rainbow trout of <0.01% (Intesal, 2009). The first clinical case of enteric redmouth disease (ERM, caused by *Yersinia ruckeri*) in Atlantic salmon fries and smolts in FW net pens was described at the end of 1992 (Troncoso et al., 1994). This case was related to increasing water temperature during the spring and summer seasons. Concurrently, during the same season a fresh water *Rickettsia* (Unidentified Agent-2, UA2) was confirmed to have caused mortalities of Atlantic salmon in lakes (Cvitanich et al., 1995). Subsequently, in June 2006 the gram-negative bacteria *Francisella* was confirmed as the agent responsible for UA2, which caused severe mortalities in Atlantic salmon farms located in Lake Llanquihue (Birkbeck et al., 2007). *Francisella philomiragia* presented high rates of morbidity and outbreaks with cumulative mortality ranging from 5% to 20% (Bohle et al., 2009). This bacterium presented 100% homology with *F. philomirhagia* subsp. *noatunensis*, currently denominated *F. noatunensis* comb. nov (Mikalsen & Colquhoun, 2010; Ottem et al., 2009).

Sea lice infestations in salmonids were first reported in Chile in 1982 (Reyes & Bravo, 1983) caused by the ectoparasite *Caligus* sp. Historically in Chile the most problematic parasite

Chilean Salmon Farming on the Horizon of Sustainability: Review of the Development of a Highly Intensive
Production, the ISA Crisis and Implemented Actions to Reconstruct a More Sustainable Aquaculture Industry

229

infection in salmonids reared in SW has been caused by *C. rogercresseyi*. The monitoring program of Intesal reported an average value of 3.38 adult *Caligus* per fish between 1999 and 2002. However, an increasing load between 2004 and 2007 was observed with an average of 3, 5 and 10 *Caligus* per Coho salmon, rainbow trout and Atlantic salmon, respectively. Moreover, the abundance levels increased up to 29, 20 and 34 *Caligus* per fish in the same mentioned species in 2007 (Rozas & Asencio, 2007). In order to get relevant information and establish strategic control measures, the national fisheries service –Sernapesca- established the Official Monitoring Program focused in marine and estuarine areas (Sernapesca, 2009). After this program was established and executed, the average load has significantly dropped (Sernapesca, 2010). However, the lower number of fish in sea sites may also explain this decrease in parasite loads after the infectious salmon anaemia event in 2007.

Between 1993 and 1994, new pathogens were detected, such as the protozoan *Nucleospora salmonis* (Bravo, 1995), which clinically affected certain strains of Atlantic salmon in SW; the metazoan *Kudoa* sp. has been sporadically diagnosed in Atlantic salmon, and the protozoan *Hexamita*, has been described in trout and Atlantic salmon fries in FW (Bluth et al., 2003). In 1995, the presence of an atypical form of *Aeromonas salmonicida* causing mortalities in Atlantic salmon in SW was confirmed, but was restricted to only a few areas in Los Lagos region (Bravo, 1999, 2000). However, in recent years, this bacterium has been diagnosed in other locations of the country and spread widely over many host species, and reported from FW sites (Godoy et al., 2010). The infection is restricted to Atlantic salmon reared in FW. It has been more frequently registered in lakes, usually after stress events. Infections are common in broodstock kept in FW from areas where the infection is endemic. There have been cases in which infection with atypical *A. salmonicida* is considered secondary in relation to other infections by bacterial agents such as *Flavobacterium sp.* or *Francisella sp.* and viral such as Infectious Pancreatic Necrosis virus, IPNv (Godoy, 2009b).

Among the emerging diseases of bacterial etiology with economic impacts in salmonid farming, *Vibrio ordalli* (Colquhoun et al., 2004) and *V. anguillarum* (Avendaño-Herrera et al., 2007) are the most important. In Chile, vibriosis occurs mainly in estuarine and sea sites, resulting in cumulative mortality up to 20%. Since 2003, vibriosis outbreaks due to *V. ordalli* have been occurring in Chile affecting initially Atlantic salmon (Colquhoun, 2004) and subsequently rainbow trout and Coho salmon.

The first cases of Amoebic Gill Disease (AGD) were reported in Atlantic salmon in 2007 (Bustos et al., 2011; Rozas, 2011). The pathogen agent in Chile is *Neoparamoeba sp.* with 99.61% of relatedness with *N. perurans* (Bustos et al., 2011; Rozas, 2011). Clinically affected fish show white patches over the gill arches, excess of mucus and bad general condition. High salinity due to low rainfall is associated to higher prevalence of the disease. Farm sites reported with AGD infection showed two times higher risk of ISA outbreaks (Rozas, 2011). Before 1997, Chilean salmon aquaculture was hit primarily by bacterial diseases. However, by mid-1998 the presence of a new disease in Atlantic salmon stock was confirmed, known as viral Infectious Pancreatic Necrosis (IPNv). In Chile, IPNv was first isolated from rainbow trout with no clinical signs (McAllister & Reyes, 1984). This event was the first report in South America, suggesting that IPNv was introduced through imported eggs from North America (McAllister & Reyes, 1984). The virus is widely distributed in the different areas where salmon farming takes place. IPNv affects Atlantic salmon, and cases in Coho salmon are considered sporadic; however, in recent years the frequency and virulence of the pathogen have increased, as well as broadened the geographical distribution. Cases have also been reported in FW recirculation systems as well as in open-flow and estuary. The disease shows high prevalence increasing from 19% in 2006 to 31.96% of total diagnoses in the year 2007 (Sernapesca, 2008a). In 2008, the monthly mortality average associated with IPNv in Coho Salmon was below 0.01%, in Atlantic salmon 0.01% and in rainbow trout 0.03% (Intesal, 2009). In 1999, ISA virus (ISAv) was first detected in SW farmed Coho salmon in Chile (Kibenge et al., 2001). In contrast to the classical presentation of ISAv in Atlantic salmon, the presence of ISAv in Chile was associated with a clinical condition characterized by jaundice in Coho salmon (Smith et al., 2002, 2006) and the virus isolation was sporadically and unsuccessful. During the winter of 2007, unexplained mortalities occurred in market-size Atlantic salmon in a grow-out site located in Chiloé (Sernapesca, 2008b). The outbreak was caused by a virulent variant of ISAv different from common European ISAv isolates. The clinical signs and lesions were consistent with the classical descriptions of the disease in marine farmed Atlantic salmon in Europe and North America (Godoy et al., 2008). The re-emergence of ISAv in Chile has resulted in one of the largest ISAv epidemics reported in the world (Mardones et al., 2009).

Upon analyzing the susceptibility of Chilean farmed salmon to the mentioned diseases, there are important differences in pathogen prevalence as well as associated losses between the three cultivated species. In terms of mortality rates, the Coho salmon is mainly affected by piscirickettsiosis during SW phase, and secondly by BKD, especially in the southern areas. The Atlantic salmon is affected mainly by IPN in FW and when first transferred to SW. After being established in SW the most relevant pathogen is ISAv, since it causes significant mortalities among infected groups (Subpesca, 2010) and secondly by piscirickettsiosis. In some cases, related to specific environmental conditions in certain geographic areas, there are important losses of Atlantic salmon due to *Caligus* sp. and atypical *A. salmonicida* and other bacterial diseases. The impact of diseases may significantly differ among farming companies, cultured species, farm sites and geographic locations.

3.2 Commercial vaccines in Chile

Nowadays 44 commercial vaccine formulations have been authorized for the use in salmonids in Chile (Servicio Agrícola y Ganadero, SAG, 2011). In Chile, the use of vaccines began in the early 1980´s, with vaccines against vibriosis (Bravo & Midtlyng, 2007). Then, after the occurrence of ERM in 1992, vaccines to protect Atlantic salmon came into use (Bravo, 1993). In 2010 a total of 6,4 million doses were used more than 6 times compared to the doses used in 2009 (Sernapesca, 2011c, 2011d). Vaccines against IPNv in Chile were rapidly available, because they were already in use in salmon farming countries. Since 2002, bivalent vaccines came into the market and were used to provide protection against IPNv and atypical furunculosis or IPNv and SRS. Currently, bivalent, trivalent, quadruple and quintuple formulations are available from international and local laboratories (SAG, 2011).

More than 10 years have passed since the first vaccine against *P. salmonis* was launched in the Chilean market. The vaccines against this bacterium apparently have no effect on reducing mortalities (Bravo et al., 2005). As of today, 26 vaccines against *P. salmonis*, both monovalent and polyvalent, have been registered in Chile (SAG, 2011), and amount of doses have increased since 2005 (Table 4). Spite the use of vaccines, piscirickettsiosis still remains a major sanitary threat for salmon farming in Chile. After the crisis caused by ISA virus, the

Chilean Salmon Farming on the Horizon of Sustainability: Review of the Development of a Highly Intensive
Production, the ISA Crisis and Implemented Actions to Reconstruct a More Sustainable Aquaculture Industry

231

use of vaccines to prevent the disease was necessary (Table 4). Thus, in April 2009, a Chilean pharmaceutical laboratory obtained a provisional registration to commercialize the first vaccine against ISAv, which has been sold since June of that year. There are currently 12 commercial vaccines registered to be used against ISAv in the domestic market (SAG, 2011).

Year	2005	2006	2007	2008	2009	2010
Disease						
IPN	38,390,234	48,639,681	16,931,232	93,356,750	145,140,655	379,817,821
IPN,SRS	248,769	-	-	18,889,551	35,086,897	61,977,044
IPN,SRS, Vibriosis	251,539	-	10,524,627	36,141,507	53,851,414	74,438,348
IPN, SRS, Vibriosis, Atypical furunculosis	-	-	16,704,008	46,689,055	19,684,380	13,106,428
IPN-Atypical furunculosis	7,840,849	4,590,883	-	-	-	-
IPN, Vibriosis	64,732,058	64,423,621	25,791,390	20,395,269	1,282,166	40,880,625
IPN, Vibriosis, Atypical furunculosis	31,762,691	23,010,340	12,412,470	3,859,378	4,228,818	885,332
ISA	-	-	-	-	11,430,448	38,599,867
IPN, Vibriosis, Atypical furunculosis, SRS, ISA	-	-	-	-	-	8,036,670
IPN, Vibriosis, SRS, ISA	-	-	-	-	-	15,634,191
SRS	27,842,546	29,926,809	16,340,430	35,439,138	41,904,431	69,057,300
SRS, BKD	nd	nd	nd	nd	nd	10,681,601
BKD	nd	nd	nd	nd	nd	770,444
Yersiniosis/ Columnaris disease	54,400,781	18,883,786	7,037,977	30,255,690	1,333,973	6,400,000*
Other	505,136	-	4,814,978	2,727,235	14,545,613	1,008,187

nd: no data available; *only doses against Yersinia

Table 4. Number of vaccine doses used in Chile during 2005-2010.

3.3 Antibiotics and chemotherapeutants in Chilean aquaculture

Within the national policy regarding the use of antibiotics, procedures for control and prevention of diseases are established. Nevertheless, these measures are general and there is a legislative gap in this area. Currently, with the amendments of the regulatory framework a number of changes in this area will be established. A program of surveillance of good practices in the use of antibiotics is being generated, aiming to improve antibiotic use in the country and to detect patterns in the use of these compounds (San Martin, 2010). In relation to the use of antibiotics in Chile, a decline in the use of quinolones and fluoroquinolones (oxolinic acid and flumequine, respectively), and an increase in the use of florfenicol has been observed in recent years, Table 5 (San Martin et al., 2010).

During the years 2005-2009, the use of other antibiotics have been relatively stable, with significant use of oxytetracycline, and a marginal share of sulfa/trimethoprim, and amoxicillin (San Martin et al., 2010). The use of florfenicol, represented over 61% of the total of antibiotics used in 2009. According to Sernapesca, piscirickettsiosis accounted for the highest diagnostic number, using over 65% of the therapeutic stock. In Chile, there are 6 antibiotics approved for use in aquaculture, this narrows the options when treatment is the choice. It is known that the rotation of antibiotics is a low frequency practice in the Chilean salmon industry, which over

Year	2005	2006	2007	2008	2009	2010
Active principle						
Oxolinic acid (kg)	50,713	39,035	74,582	25,325	2,900	1,192
Amoxicillin (kg)	97	253	1,732	349	473	863
Erythromycin (kg)	1,994	1,972	2,139	7,981	1,542	2,586
Florfenicol (kg)	33,258	102,838	143,009	184,715	113,137	74,431
Flumequine (kg)	96,751	95,575	74,773	32,293	3,233	1,588
Oxitetracycline (kg)	56,354	104,135	89,309	74,931	63,172	62,506
Sulfa+trimethoprin (kg)	0	1	91	22	11	-
Total (kg)	239,167	343,808	385,635	325,617	184,469	143,165

Table 5. Amount of antibiotics used by the Chilean salmon industry, by active principle and year.

the course of the years has led to the emergence of resistant organisms (San Martin et al., 2010). Due to the low number of antibiotics approved, it is suggested to register new drugs in the country, or to allow the extra-label use of antibiotics, because the latter would allow veterinarians to use other drugs that are not registered for salmon but within the legal framework. However, this process is not authorized (San Martin et al., 2010).

Several chemotherapeutants have been used to control caligidosis in Chile, e.g. metriphonate (Neguvon™) was the first product used to control *Caligus teres* between 1981 and 1985, replaced by another organophosphate, diclorvos (Nuvan™) between 1985 and 2000, both used as bath treatment. Ivermectin use in feeds was introduced in Chile at the end of the 1980s, but between 2000 and 2007 the only veterinary product authorized for this use in Chile was the avermectin (emamectin benzoate EMB), without alternative treatments available. When EMB (Slice™) was introduced into the Chilean market in 1999, one standard treatment with the product (50 µg of active ingredient per kilogram biomass daily for 7 day) controlled the sea lice infestations on the salmon for at least five weeks in the summer and even longer in the winter (Bravo, 2003). Since early 2005, a notable loss of efficacy was observed in several fish farms, resistance development was the main cause (Bravo et al., 2008). Within the management of this parasitic disease, a strategy of coordinated treatments was established by the authority, which aimed to reduce the infestation levels over subzones or groups of concessions. Today, there are 10 commercial products for caligidosis treatment authorized by SAG. The current compounds for treatments are deltametrine 1%, diflubenzuron 80% and emamectin benzoate 0.2% (Table 6). The last antiparasitic drug authorized was cipermetrin 5% (Betamax) (SAG register No. 2085, 2010).

4. Sanitary crisis of the Chilean salmon industry

The ISAv crisis may not be exclusively related to the rapid spread of a highly virulent pathogen. Prior to the ISAv outbreak, the industry struggled with serious caligidosis and piscirickettsiosis outbreaks, especially in Los Lagos region (X). In addition, several factors affecting the quality of the smolt has not been fully tackled such as excessive handling, grading, varying FW quality, high density, and water quality and welfare in long distances on land and sea transport. Perhaps one of the most threatening practices in Chilean salmon farming that relate to spreading of diseases was the inadequate disinfection management of wastes from processing facilities. Additionally, boat travel between the farming facilities

Year	2005	2006	2007	2008	2009	2010
Active principle						
Emamectin benzoate (kg)	240	443	595	285	65	47
Diflubenzuron (kg)	0	0	0	162	3878	3639
Deltamethrin (1%)*	0	0	516	10524	3168	3431
Cypermethrin (5%)*	-	-	-	-	-	593

*values are given in litres

Table 6. Antiparasitic drugs used through feed or bath treatments.

may have contributed to the risk of disease transfer in Chile. In spring of 2007, barely four months after the detection of ISAv in two sites around Chiloé Island, the number of infected sites had tripled (Carvajal, 2009). Nevertheless, the industry was optimistic and considering that as an isolated event limited to one farming company, and there were little sense that this could be a catastrophic event. Although the industry argued that the media was blowing the situation out of proportion, a few months later ISA had reached all salmon producing regions in Chile, affecting almost every salmon farming company. At this point, the need of a consorted action between the authorities and the industry was evident. The disease began to spread in late December 2007. Outbreaks were reported in Aysén region and ISAv positive fish in different sea sites and broodstock were reported late in 2008. Between July 2007 and July 2008, 74 sea sites, presented positive results to virus detection, out of which 44 were classified as focal outbreaks. Among the positive farms, 89% were located in Los Lagos region and the remaining 11% in Aysén region. Furthermore, the impacts of this disease account for the elimination of >11,000 tons of fish distributed in 250 sites which belonged to 13 different companies. The ISA outbreak reached its peak in 2008 affecting 93 sites (Silva, 2010b). However, there was a progressive decrease in outbreak reporting since 2008, with 3 outbreaks in late 2009 and 4 outbreaks in 2010. This decline is explained by the decrease in active sites, where the industry capacity of 550 authorized sites was operating in only its 20%. It can also be noted that the area with most outbreaks was Central and Southern Chiloé where barely 180 sites were operating during 2009 (Silva, 2010b). As a consequence of ISA, Chile stopped earning more than USD 883 million in 2008 (Silva, 2010b). In 2009, the industry export volumes fell by 16% and earnings experienced a decrease of 12% compared with the previous year (Silva, 2010c). When ISA outbreaks started, the smolts transferred in to sea sites decreased in about 600 thousand a month. In 2006 about 12 million smolts were transferred per month. The smolt release in 2009 was reported to be less than the tenth of what was in 2007 (Asche et al., 2010), and a fall of 83% in January 2009 when compared with the same period in 2008. In mid-2009 the situation began to improve, and in mid-2011, every 30 days, four million smolts were transferred and the figure is increasing. This is due to changes in the production model, which has reduced mortality. During January 2010, the number of fish in farms was 61% of what was in 2008. During 2010, the exports reached 352,637 tons a 23% less than the previous year, showing a variation in earnings of only - 2% (TechnoPress, 2010). As of July 2011, about 20 sites have been reported as ISAv HPR0 positive, a low-pathogenic variant, only one site has been confirmed to remain in outbreak due to a pathogenic strain (HPR2). This number of infected sites well correlates with increasing aquaculture activities after the crisis.

5. Aquaculture governance

During the 1980s, the incipient salmon aquaculture was ruled by a supreme decree for the practice of fishing, which was clearly inadequate for the forthcoming rapid growth of the sector. However; since 1991, the regulation towards any aquaculture activity in Chile has been regulated by the General Law of Fisheries and Aquaculture (Ley General de Pesca y Acuicultura, LGPA, Law No.18.892). This law also regulates fisheries and conservation of living resources, industrial and small-scale capture, scientific and recreational fisheries. In addition, some articles regard commercialization, processing, storage and transportation of fishery products. The main administrative authority responsible for aquaculture and fisheries is the Ministry of Economy, Development and Reconstruction. Within this ministry, the Subpesca provides all the information and technical support required to allow the Minister to take actions and measures towards aquaculture and fishery activities. In addition, functional matters, including enforcement, are carried out by Sernapesca. Moreover, the Undersecretary of Marine Affairs, the Environmental Commission (CONAMA), the Directorate of Boundaries and State Limits, CORFO, the Land Registry and the Directorate General for Water intervene in some of the procedures for the granting of concessions and other authorizations within aquaculture.

5.1 Aquaculture regulatory framework (1991-2010)

The enactment of the LGPA aimed to organize the responsibilities of authorities and delegates to dictate the regulations pertaining to the sustainable operation of aquaculture. The law provided Subpesca an improved organization in order to expedite its surveillance and faculties. During the 1990s, there were several Supreme Decrees (S.D.) implemented that were relevant for aquaculture such as the S.D. No.475 and S.D. No.499, clarifying national policy for the use of Chile's littoral coastline and national register of aquaculture in 1994, and the regulation on information of fishery and aquaculture activities by S.D. No.464 in 1995. In addition, in 1993, the S.D. No.550 and No.290 stated regulations towards site´s size, authorization, concessions and inscription in aquaculture activities such as portion of water used according to the area, kind of system and type of organism. Furthermore, the general basis for environmental regulation (Law No.19,300) established the national environmental commission, CONAMA, and introduced the obligatory environmental impact evaluation for any productive activity, including aquaculture. CONAMA and later the regional (COREMA) are responsible for the environmental management and supervision. Regular environmental controls and level for contaminants and emissions, but also quality of water, air and soil are established, as well preventive management, environmental programs and decontamination. This law also establishes punishment for environmental damage. Implementation of the system for evaluation of environmental impact SEIA, S.D. No.30 in 1997 and S.D. No.95. in 2001, that obliges to provide a study of environmental impact or environmental impact assessment to any investment project, which could cause impacts according to the Law. Therefore, the LGPA has been actively amended since intensive and extensive aquaculture begun in Chile.

5.2 Environmental and sanitary regulations

The environmental regulation for aquaculture RAMA, S.D. No.320 in 2001, states the environmental requirements for approval of aquaculture activities. It demands that net changing, washing and antifouling treatment must be conducted in inland sites, using water treatments. It also introduced the preliminary site characterization (CPS) for any inland or marine sites, demanding an environmental impact assessment, which improves the evaluation process. Furthermore, it establishes annually environmental monitoring as part of the environmental information program (INFA). Operating sites should annually present an INFA-report with the impact on the oxygen conditions at the bottom below the cages. This is of especial relevance for remaining sites located in lakes. When anaerobic condition is stated, the production can be on hold for one year, if the anaerobic condition is kept, the site should reduce 30% of the production and consecutively for the following years when the condition is not improved. Both CPS and aerobic conditions will be surveilled by competent persons or companies approved by the authority. In 2003, a score system for the sites based in type, production size, bathymetry and bottom conditions was introduced. In addition, the methodology for analysis of CPS and INFA according to the score was established in 2006. Various measurements were introduced towards management for solid and liquid residues, for harvesting, mortality and processing plants in 2008.

The sanitary regulation for aquaculture RESA, S.D. No.319 in 2001, stated that a sanitary condition depends on the prevention and control of events. This regulation was enacted to prevent and control high risk diseases in aquatic species securing sanitary controls, surveillance and eradication of infectious diseases in farmed fish within the country. Sernapesca was provided with larger resources to cover the increased inspection needs. It established a classification of the high risk illness, basic conditions for production sites, experimental sites, processing plants, transport, waiting sites, harvesting, egg´s import, domestic egg´s production, certifications, also conditions and task of the reference and diagnostic laboratories. It also established that only registered pharmaceutical products for aquatic species might be used. This regulation was recently updated (September 2011) including major pillars regarding pathogens such as bioexclusion, biocontainment measures and sanitary management as well as reinforcement of preventive measures in FW systems.

5.3 Specific sanitary programs and regulations

Since 2003, the general sanitary programs have been implemented through 12 resolutions (Rs. 60 to Rs. 72) by Sernapesca to regulate sanitary issues in aquaculture. The Rs. 60 regulates the registration, proceedings and conditions for vaccination of farmed fish. The Rs. 61 to 63 established proceedings towards information, data processing, reporting, surveillance and actions to be taken when detection and outbreak of high risk diseases (HRD list 1 and 2, Table 7), but also for unknown aetiology and non-described diseases in a specific zone. Farm sites and diagnostic laboratories may be evaluated towards diseases of HRD list 2 and confirm absence of HRD list 1, at least twice a year.

These resolutions tackle transport of fish, eggs, high risk biological material, feed, equipment, and materials in any farming activity according with the zoning, also proceedings for disinfection and traceability applying for domestic and imported eggs, conditions, procedures and management for mortality such as frequency and methods for mortality check, necropsy, and disinfections. In order to prevent outbreaks, every farm should have procedures for treatments and handouts for a correct medication, which is also ruled by the Rs. 67. The Rs. 68 establishes the management and proceeding for all kind of waste in farming, slaughter and processing plants (liquid, solid, organic and inorganic), introducing proceedings for personal, disinfection and

List 1	List 2
Epizootic Haematopoietic Necrosis (EHN)	Infectious Pancreatic Necrosis (IPN)
Infectious Haematopoietic Necrosis (IHN)	Piscirickettsiosis
Oncorhynchus masou virus disease (OMVD)	Streptococcosis
Viral hemorrhagic septicaemia (VHS)	Infectious Salmon Anaemia (ISA)
Spring viraemia of carp (SVC)	Jaundice/Icteric syndrome
Viral encephalopathy and retinopathy (VER)	Atypical furunculosis
Enteric septicaemia of channel catfish (ESC)	Vibriosis
White sturgeon iridoviral disease(WSIVD)	Caligidosis
Furunculosis	Amoebic gill disease (AGD)
Epizootic ulcerative syndrome (EUS)	Renibacteriosis (BKD)
Gyrodactylosis	
Pancreas Disease (PD)	
Red sea bream iridoviral disease (RSIVD)	

Table 7. Classification of High Risk Diseases (Rs. 2352, 2008, Subpesca).

treatments. The Rs. 70 introduces procedures for the sanitary control of broodstock and general sanitary conditions for domestic broodstock, spawning, egg and reproduction, including physical barriers for facilities, in/out water treatment. In broodstock, an individual sanitary control for HRD list 2 including PD and BKD should be performed. The Rs. 71 or sanitary management for feed establishes the use of medicated feed (personal, storage, traceability, corrective actions, monitoring proceedings, etc.). The Rs. 72 provides cleaning and disinfection procedures in fish production, hence establishes conditions for proceeding towards cleaning and disinfection in fish farming facilities applying for facilities, tools, equipment, personal, working clothes, and water. It also applies for vessels, trucks, wellboats, etc. Furthermore, specific programs regarding surveillance and control of diseases in farmed fish have been launched by Sernapesca since 2004. The specific program for control and surveillance of caligidosis established proceedings and obligatory managements for sea and estuarine water production including treatments, reporting data, and others. The specific sanitary program for the control of Caligus was enacted in 2008. Further, in 2010 the Rs. No. 873 strengthens the control and certification for the origin of imported egg from countries where PD and ISA are present. The resolution demands individual screening for the broodstock, especially for keeping the country free of the HRD list 1 (Table 7). The Specific Surveillance and Control Program for ISAv went into effect in 2008. Currently, a new regulation came into force in 2011, although having the same focus as the former in 2008; it is stricter in immediate notification and relies in a risk-based surveillance.

6. Restructuration of the industry and new legislations

The latest amendment of the LGPA was performed in 2010 driven by the severe social and economic consequences of the sanitary and production crisis. Although the draft of amendment for the LGPA was developed during first quarter of 2009, it was not until March of 2010 that it was enacted. This amendment has been the largest and most significant for the Chilean aquaculture. Some of the essential issues in this law are attribution to establish appropriate areas for aquaculture activities (AAA), which is a place with common epidemiologic, oceanographic, operational or geographic characteristics. It also has complementary environmental and sanitary regulations and procedures for licence and concession's sites. It contains reliable information regarding alternative uses of the site such as natural reserve or protected, touristic, indigenous population, harbour conflicts, hydro biological resource and especially natural banks. Since 1991, a concession site was granted for indefinable time where nowadays it is established for 25 years and renewable. In addition, it considers development of farming density regulation, strengthens regulations towards waiting and slaughter centres, zoning of the coastline border and strengthens the regulation regarding egg imports. Furthermore, the cost of an aquaculture concession has been increased from around 26 USD /ha, in 1991, to 80 USD /ha, in 2010. One of the most important changes set by the new amendments is the ban on fish movements from and between sea sites that affects the movement of broodstock from sea sites to FW facilities as well as the temporary use of estuarine sites. Several companies have begun to secure their broodstocks in FW facilities and to perform whole cycle in FW. Interestingly, these amendments also include the concept of animal welfare where aquaculture production should carry out procedures that avoid unnecessary suffer of animals; however, the criteria for welfare have not been established. In general, it should be considered by transport, design, farm density, treatment, light exposition, among others.

6.1 Aquaculture neighbourhoods

The LGPA states that concessions should be relocated within an AAA; hence, the distribution of several sites should be modified and relocated. Thus, within the farms in an AAA, coordinated actions should be carried out as fish inputs, prophylaxis and therapeutic treatments, sanitary issues, harvest, and resting periods. Each concession will hold one vote to agreements within the "neighbourhood" which should be approved, public published and carried out according the regulations by Subpesca. Therefore, there have been major challenges in the relocation of salmonid farms in neighbourhoods in the extensively used areas within Los Lagos and Aysén region.

The zoning of Los Lagos and Aysén regions was established by the Rs. No.450 in 2009. It states several suitable zones for aquaculture activities following the oceanographic, epidemiologic and operative, logistic characteristic towards ISAv infection and control. In 2010, it was established a total of 24 aquaculture producing neighbourhoods in Los Lagos region and 56 in Aysén region (Figure 8). While in Magallanes region there is still no established geographic zoning of the coastal border, and neither neighbourhoods nor new concession can be granted. But, it should be defined by 2011. A discussion and process that are currently carried out by local fishery associations, tourism entities, mining companies, aquaculture producers, NGOs and regional government with the support of the congress commission for fisheries. The distance between neighbourhoods have been established in minimum of 3 nautical miles (ca. 5.6 km) and aquaculture sites must be spaced out at least in 1.5 nautical miles (ca. 2.8 km) from marine protected areas (natural park and reserve) and from another site. When a land protected area has a sea border, the coastal zoning should establish a secure distance to the closest aquaculture site.

6.2 Agreements and certifications

Unlike the mandatory government policies used in other countries, Chile has promoted voluntary implementation of traceability and quality assurance systems. This has allowed

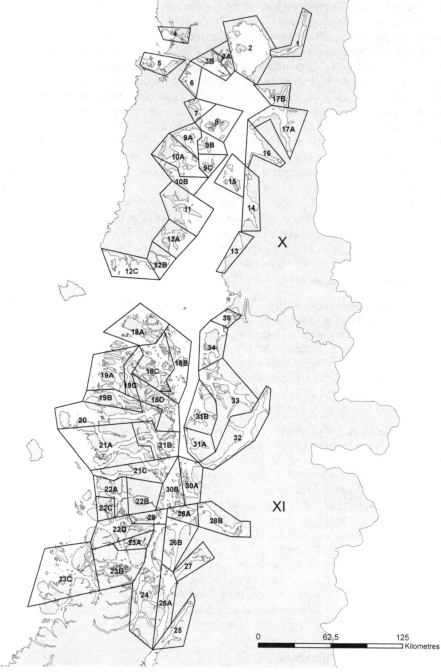

Fig. 8. Salmonid aquaculture neighbourhoods in Los Lagos region (X) and Aysén region (XI). (map source: GeoAustralis).

Chilean Salmon Farming on the Horizon of Sustainability: Review of the Development of a Highly Intensive
Production, the ISA Crisis and Implemented Actions to Reconstruct a More Sustainable Aquaculture Industry

239

the private sector time to plan for and adapt to export requirements and develop systems without imposed financial pressure or deadlines. Intesal released sanitary measurements for the associates in December 2009, which was further improved and re-edited in August of 2010. The effects of these actions may be confirmed subsequently in the next 5 years. It proposes 44 measurements, whereas 20 are complementary to the new resolutions and amendments of the LGPA, 6 are partially a part of the regulation and 18 are part of regulation (Intesal, 2010). To a large degree, Chile has allowed large export markets to define traceability requirements for Chilean products. In response to emerging international requirements for traceability, SalmonChile facilitated the development of an Integrated Management System (SIGES) in 2003. In order to obtain the SIGES certification, companies must commit to adhere to otherwise voluntary protocols (e.g., information system software, ISO, OSHA, HACCP, best practice guidelines, etc.). The SIGES is a voluntary agreement for the associated companies that implement standards of quality, welfare, clean production, environment, and working health and security, recognized by domestic and international regulations. It facilitates the implementation of ISO 9,001, ISO 14,001 and OHSAS 18,001. Although in 2006, the worldwide biggest food-market chain, Wal Mart (USA), recognized SIGES-SalmonChile and Safe quality Food (SQF-SIGES) it is still not recognized by shareholders worldwide.

In October 2010, SalmonChile signed an agreement with GLOBALG.A.P. to launch standards for salmon production also called SALMONG.A.P., which is the first internationally recognized standard for the cultured salmon in Chile. SALMONG.A.P. aims improvement of salmon production and processing towards best practices, which considers multi-criteria task such as feed, health, safety, quality, environmental issues, working conditions, welfare and biosecurity.

In 2000, a special programme for public-private association was created by CORFO in the field of cleaner production, operated by the National Clean Development Council. Clean production agreements (APL) have strategies for production and environmental management, which are voluntarily joined by industry and government partners; however, once an APL is signed the different tasks and steps should be reached and in case of non-fulfilment, the authority may punish partners according to the regulation. In 2010, producers and Intesal improved their production practices by the first APL within Los Lagos region, ratified by all the members of SalmonChile, 48 companies at that time, representing one of the most important public-private agreements towards sustainable aquaculture. This agreement covers 46 action points that have been carried out. The second APL within the industry was ratified by FW producers of La Araucanía region where 28 FW producers signed an APL in 2010.

6.3 The role of the NGOs

Similarly to other salmon producing countries several non-governmental organizations (NGOs) have watched all the impacts of the industrial aquaculture and published several reports and launched campaigns regarding aquaculture's negative effects on the local and global environment as well in the communities and social issues. In Chile, especially since 2000s several NGOs have being active such as Terram, Olach, Oxam, Oceana and WWF. Nevertheless, WWF has published a number of reports regarding impacts of the Chilean aquaculture. These promote dialogue between NGOs, community, producers and public entities suggesting solutions as aquaculture standards. Lately, WWF Chile has published 5 reports regarding the effect and assessment of the FW production, environmental issues, salmonid escapes, and carrying capacity of fjords (Díaz & León-Muñoz et al., 2006; León-Muñoz et al., 2007; Nieto et al., 2010; Sepúlveda et al., 2009; Tapia & Giglio, 2010).

7. Perspectives

The experience from all fish producing countries including Chile has demonstrated that the fish health situation plays a major role when expecting good biological and economical performance in aquaculture farming. On a larger level, this aspect is determinant to success or failure of an industry focused on the artificial farming of fish (Smith et al., 2001). Because of the rapid aquaculture expansion, diseases have emerged with disastrous economic consequences. Estimated disease losses in aquaculture worldwide are in the order of USD 8 billion per year, which represents 15% of the value generated by the global aquaculture production (Enright, 2003). The economic losses attributed to diseases are mostly attributed by mortality caused directly by diseases; although weight losses, quality losses and costs incurred because of control and prevention are also important cost drivers (Bravo et al. 2005). Severe sanitary events occurred in salmon producing countries in the early 1990´s; in Norway, and in the middle and late 1990´s in Scotland, Faroe Island, Canada, USA and Chile in the 2000´s (Gustafson et al., 2005; Lynstad et al., 2008). It has been clearly demonstrated that in Chile the regulatory system was ineffective by reducing environmental and sanitary threats prior to the ISA crisis in 2007, despite RAMA and RESA. Furthermore, the voluntary agreements (APL) were either inadequate or insufficient or delayed. Thus, it is evident that the industry itself fails to implement adequate and timely measurements in response to threats. Although Chile has been the largest producer of rainbow trout and Coho salmon, and the second producer of Atlantic salmon worldwide since 1990s, it was not until 2010 that authorities and the industry developed together the designing, regulation and ruling of organized aquaculture neighbourhood with several measurements regarding mainly biosecurity and production. Although up to date about 21 sites have been reported with the low pathogenic ISAv-HPR0 strain, only 2 site has been reported as ISAv-HPR7b recording clinical disease and mortalities. In addition, in the beginning of 2011, reports from the industry indicate that the salmon production in Chile will be back to 2007 level in 2013. The fast comeback has been possible due to willingness from the banks to reorganize the debts of the companies, extremely high salmon prices and risk willing investors. In Chile, research and development is needed to secure a sustainable development of the industry. The sanitary crisis, caused by an industry that were allowed to grow too fast considering the unsolved disease problems, e.g. SRS, *Caligus*, IPNv and finally the epidemic ISAv, has shown that Chile has the obligation as a country to stimulate the industry to invest more resources into R&D. Clearly, Chilean challenges require local solutions and there cannot be good regulation or good production models without a sound scientific and technical rationale. Historically in Chile, the public investment in research and development in aquaculture yearly, which involves private partners, has been at average below USD 10 per ton of farmed fish (Medina, 2008 pers. Comm.). Most of these resources have been used to implement knowledge developed in other countries, but also to build up a research infrastructure for aquaculture at Universities. More recently, also international research institutes as Nofima, Sintef, VESO and NIVA have established in Chile. Further, several farming and feeding companies are building up their own R&D departments in Chile with the aim to develop a more sustainable industry. Altogether, we have an optimistic view on the future of Chilean salmon farming.

Chilean Salmon Farming on the Horizon of Sustainability: Review of the Development of a Highly Intensive
Production, the ISA Crisis and Implemented Actions to Reconstruct a More Sustainable Aquaculture Industry

241

8. Conclusions

In the 2000s the growing projections in Chilean aquaculture were very optimistic, but Chilean salmon production has not increased according to these expectations in the last four years. The major set-back is caused by severe sanitary events appearing at the same time. Specifically, permanent losses due to piscirickettsiosis (SRS) as well as the spread of treatment-resistant *Caligus* and lately the occurrence of epidemic ISAv have presented sanitary challenges of a magnitude that surprised the industry and rendered earlier growth projections obsolete. The sanitary crisis that led to economical, environmental and social problems, forced the authorities to improve and strengthen aquaculture regulations, providing more governmental control. In addition, the industries' own organisation, SalmonChile, through Intesal has established several measures towards sanitary, production, and logistics management that complements the new governmental regulations, where an essential issue is the fact that neighbourhood will operate in an organized manner. Although the industry are facing increased production cost, these new regulations and measurements are massively accepted as the unique way to reborn the Chilean salmonid aquaculture.

9. Acknowledgment

The authors acknowledge the valuable information provided by TechnoPress, Intesal and GeoAustralis. The authors also thank Luis Martinez for the assistance in information search for this review.

10. References

Águila, M.P. & Silva, G (2008). Salmonicultura en Lagos: En la Senda de la Sustentabilidad, *Aqua Acuicultura + Pesca*, Vol.20, No.120, pp 8-14

Asche, F.; Hansen, H.; Tveterås, R. & Tveterås, S. (2010). The Salmon Disease Crisis in Chile. *Marine Resource Economics*, Vol.24, No.4, pp. 405-411, ISSN 0738-1360

Avendaño-Herrera, R.; Silva-Rubio, A. & Toranzo, A. E. (2007). First Report of Vibrio *anguillarum* in chilean Salmon Farms. *American Fisheries Society/Fish Health Section Newsletter*, Vol.35, No.2, pp. 8-9

Banco Central de Chile (2010). Estadísticas Económicas. In: *Series de Indicadores*, 27.05.2011, Available from
http://www.bcentral.cl/estadisticas-economicas/series-indicadores/index.htm.

Barton, J. R. & Murray W. E. (2009). Grounding Geographies of Economic Globalisation: Globalised Spaces in Chile's Non-Traditional Export Sector, 1980–2005. *Tijdschrift voor economische en sociale geografie*, Vol.100, No.1, pp. 81-100, ISSN 1467-9663

Barton, J. R. (1997). Environment, sustainability and regulation in commercial aquaculture: The case of Chilean salmonid production. *Geoforum* Vol.28, No.3-4, pp. 313-328, ISSN 0016-7185

Birkbeck, T. H.; Bordevik, M.; Froystad, M. K. & Baklien, A. (2007). Identification of *Francisella* sp from Atlantic salmon, *Salmo salar* L., in Chile. *Journal of Fish Diseases*, Vol.30. No.8, pp. 505-507, ISSN 1365-2761

Bjørndal, T. (2002). The competitiveness of the Chilean salmon aquaculture industry. *Aquaculture Economics & Management*, Vol.6, No.1, pp 97-116, ISSN 1365-7305

Bluth, A.; Espinosa, L.; Guzmán, C.; Hidalgo, C.; Martínez, S.; Mellado, M.; Sánchez, V. & Tang, M. (2003) *Aquaculture in Chile*, TechnoPress S. A., ISBN 956-8042-02-4, Santiago, Chile

Bohle, H.; Tapia, E.; Martínez, A.; Rozas, M.; Figueroa, A. & Bustos, P. (2009). *Francisella philomiragia*, bacteria asociada con altas mortalidades en salmones del Atlántico (*Salmo salar*) cultivados en balsas-jaulas en el lago Llanquihue. *Archivos de Medicina Veterinaria*, Vol.41, No.3, pp. 237-244, ISSN 0301-732X

Branson, E. J. & Nieto Diaz-Muñoz, D. (1991). Description of a new disease condition occurring in farmed coho salmon, *Oncorhynchus kisutch* (Walbaum), in South America. *Journal of Fish Diseases*, Vol.14, No.2, pp. 147-156, ISSN 1365-2761

Bravo, S. & Midtlyng, P. (2007). The use of fish vaccines in the Chilean salmon industry 1999-2003. *Aquaculture*, Vol.270, No.1-4, pp. 36–42, ISSN 0044-8486

Bravo, S. (1993) Diseases reported in pen reared salmonids from Chile. *American Fisheries Society/Fish Health Section Newsletter*, Vol.21, No.3, pp. 3

Bravo, S. (1995). Reporte de *Enterocytozoon salmonis* en Chile. *Chile Pesquero*, Vol.18, No.87, pp. 23-25, ISSN 0716-1956

Bravo, S. (1999). Atypical Furunculosis in Atlantic salmon. *American Fisheries Society/Fish Health Section Newsletter*, Vol.27, No.4, pp. 2

Bravo, S. (2000). Occurrence of atypical furunculosis in Chile. *Bulletin of the European Association of Fish Pathologists*, Vol.20, No.5, pp. 209–211, ISSN 0108-0288

Bravo, S. (2003). Sea lice in Chilean salmon farms. *Bulletin of the European Association of Fish Pathologists*, Vol.23, No.4, pp. 197-200, ISSN 0108-0288

Bravo, S., Campos, M. (1989a) Síndrome del Salmón Coho. *Chile Pesquero*, Vol.12, No.54, pp. 47-48, ISSN 0716-1956

Bravo, S., Campos, M. (1989b) Coho salmon syndrome in Chile. *American Fisheries Society/Fish Health Section Newsletter*, Vol.17, No.3, pp.3

Bravo, S.; Dolz, H.; Silva, M.; Lagos, T.; Millanao, C. & Urbina, A. (2005). Diagnóstico del uso de fármacos y otros productos químicos en la acuicultura. *Report Proyect N° 2003-28*. Available from
http://www.fip.cl/FIP/Archivos/pdf/informes/inffinal%202003-28.pdf

Bravo, S.; Sevatsal, S. & Horsberg, T. (2008). Sensitivity assessment of *Caligus rogecresseyi* to emamectin benzoate. *Aquaculture*, Vol.282, No.1-4, pp. 7-12, ISSN 0044-8486

Bustos, P.A.; Young, N.J.; Rozas, M.A.; Bohle, H.M.; Ildefonso, R.S.; Morrison R.N. & Nowak, B.F. (2011). Amoebic gill disease (AGD) in Atlantic salmon (*Salmo salar*) farmed in Chile. *Aquaculture*, Vol.310, No.3-4, pp. 281-288, ISSN 0044-8486.

Carvajal, P. (2009). ISA and the reshaping of Chile's salmon industry. *Industry Report*. IntraFish Media US. Norway. Available from
http://www.intrafish.no/global/industryreports/article247570.ece

Claude, M.; Oporto, J.; Ibañez, C.; Brieva, L.; Espinosa, C. & Arqueros, M. (2000). La ineficiencia de la Salmonicultura en Chile: Aspectos sociales, económicos y ambientales. In: *Fundación Terram*, Available from
http://www.terram.cl/nuevo/images/storiesrppublicos1.pdf

Colquhoun, D.J.; Aase, I.L.; Wallace, C.; Baklien, Å. & Gravningen, K. (2004). First description of *Vibrio ordalii* from Chile. *Bulletin of The European Association of Fish Pathologist*, Vol.24, No.4, pp. 185–200, ISSN 0108-0288

Cvitanich, J.; Garate O.; Silva, C.; Andrade, M.; Figueroa, C. & Smith, C. (1995). Isolation of a new Rickettsia-like Organism from Atlantic Salmon in Chile. *American Fisheries Society/Fish Health Section Newsletter*, Vol.23, No.1, pp. 3

Cvitanich, J.; Gárate, O. & Smith, C.E. (1991). The isolation of a rickettsia-like organism causing disease and mortality in Chilean salmonids and its confirmation by Koch's postulate. *Journal of Fish Diseases*, Vol.14, No.2, pp. 121-145, ISSN 1365-2761

Díaz, S. & J. León-Muñoz (2006). Synopsis of salmon farming impacts and environmental management in Chile, In: *Consultancy Technical Report*, World Wildlife Fund (Ed.), 1-88, Valdivia, Chile.

Fryer, J.; Lannan, C.; Garcés, L.; Larenas, J. & Smith P. (1990) Isolation of a rickettsiales-like organism from diseased coho salmon (*Oncorhynchus*) in Chile. *Fish Pathology*, Vol.25, No.2, pp. 107–114, ISSN 0388-788X

Fryer, J.; Lannan, C.; Giovannoni, S.J. & Wood, N.D. (1992). *Piscirickettsia salmonis* gen. nov., sp. nov., the causative agent of a epizootic disease in salmonid fishes. *International Journal of Systematic Bacteriology*, Vol.42, No.1, pp 120-126, ISSN 1466-5034

Godoy, M. (2009a). Síndrome Rickettsial Salmonídeo (SRS) en Trucha Arcoiris (*Oncorhynchus mykiss*) I: Presentación Clínica Crónica, Abscedativa. In: *http://www.marcosgodoy.com*, 3.08.2010, Available from http://www.marcosgodoy.com/foro/?p=393.

Godoy, M. (2009b). Furunculosis atípica (*Aeromonas salmonicida atípica*) en Salmón del Atlántico (*Salmo salar*) cultivado en agua dulce I: Signos clínicos macroscópicos. In: *http://www.marcosgodoy.com*, 27.05.2011, Available from http://www.marcosgodoy.com/foro/?p=136

Godoy, M.; Aedo, M.; Kibenge M.; Groman D.; Yason, C.; Grothusen, H.; Lisperguer, A.; Calbucura, M.; Avendaño, F.; Imilán, M.; Jarpa ,M. & Kibenge, F. (2008). First detection, isolation and molecular characterization of infectious salmon anaemia virus associated with clinical disease in farmed Atlantic salmon (*Salmo salar*) in Chile. *BioMed Central Veterinary Research*, Vol.4, No.28, pp. 1-13, ISSN 1746-6148

Godoy, M.; Gherardelli, V.; Heisinger, A.; Fernandez, J.; Olmos, P.; Ovalle, L.; Ilardi, P.; Avendaño-Herrera, R; (2010). First description of atypical furunculosis in freshwater farmed Atlantic salmon, *Salmo salar* L., in Chile. *Journal of Fish Diseases*, Vol.33, No.5, pp. 441–449, ISSN 1365-2761

Iizuka, M. (2004). Organizational capability and export performance: the salmon industry in Chile, Proceeding of DRUID Winter Conference, pp 1-23, Aalborg, Denmark, January 22-24, 2004

InnovaChile (2008). Desarrollo y producción de vacuna oral y de fármaco inocuo contra *Pisicirickettsia salmonis* para aumentar la competitividad y sustentabilidad de la industria salmonera . In: *Concurso nacional de proyectos de innovación de interés público e innovación precompetitiva 2007*. CORFO. 27.05.2011. Available from http://www.corfo.cl/incjs/download.aspx?glb_cod_nodo=20080701143625&hdd_nom_archivo=Proyectos_aprobados.pdf

Intesal (2009). Programa Zonal Instituto Tecnológico del Salmon, Intesal 2008. pp. 24-28.

Intesal (2010). Medidas Sanitarias de la Industria del Salmón de Chile, In: *Intesal de SalmonChile*, medidas sanitarias, documentos, medidas sanitarias y anexos, 27.05.2010, Available from http://www.intesal.cl/

Kibenge, F. S.; Garate, O. N.; Johnson, G.; Arriagada, R.; Kibenge, M. J. & Wadowska, D. (2001). Isolation and identification of infectious salmon anemia virus (ISAV) from coho salmon in Chile. *Diseases of Aquatic Organisms*, Vol.45, No.1, pp. 9-18, ISSN 1616-1580

Larenas, J.; Contreras, J. & Smith, P. (2000) Piscirickettsiosis: Uno de los principales problemas en cultivos de salmones en Chile. *Tecno Vet*, Vol.6, No.2, ISSN 0718-1817

León-Muñoz, J., D. Tecklin, A. Farías and S. Díaz (2007). Salmon farming in the lakes of Southern Chile - Valdivian Ecoregion - History, tendencies and environmental impacts. In: *History, Tendencies and Environmental Impacts*, World Wildlife Fund (Ed.), 1-44, Valdivia, Chile.

Mardones, F.; Perez, A. & Carpenter, T. (2009). Epidemiologic investigation of the re-emergence of infectious salmon anemia virus in Chile. *Diseases of Aquatic Organisms*, Vol.84, No.2, pp. 105–114, ISSN 1616-1580

McAllister, P. & Reyes, X. (1984) Infectious pancreatic necrosis virus: isolation from rainbow trout *Salmo gairdnieri* Richardso, imported into Chile. *Journal of Fish Diseases*, Vol.7, No.4, pp. 319–322, ISSN 1365-2761

Mikalsen, J. & Colquhoun, D.J. (2010). *Francisella asiatica sp. nov.* isolated from farmed tilapia (*Oreochromis sp.*) and elevation of *Francisella philomiragia subsp. noatunensis* to species rank as *Francisella noatunensis comb. nov., sp. nov. International Journal of Systematic and Evolutiuonary Microbiology* doi:10.1099/ijs.0.002139-0. ISSN 1466-5026

Nieto, D., R. Norambuena, E. González, L. González and D. Brett (2010). Smolt production systems in Chile. In: *Analysis of alternatives from the environmental, sanitary and economic perspectives*, World Wildlife Fund (Ed.), 1-48, Valdivia, Chile.

Norambuena, R. & González, L. (2006). National Aquaculture Sector Overview: Chile, In: FAO Fisheries and Aquaculture Department, 27.05.2011, Available from http://www.fao.org/fishery/countrysector/naso_chile/en#tcN900AF

Oliva, D.; Sielfeld, W.; Durán, L. R.; Sepúlveda, M.; Pérez, M. J.; Rodríguez, L.; Stotz, W. & Araos, V. (2003). Interferencia de mamíferos marinos con actividades pesqueras y de acuicultura, *Report of Proyecto FIP 2003-32*, Available from http://www.fip.cl/FIP/Archivos/pdf/informes/inffinal%202003-32.pdf

Olivares, J. & Marshall, S. (2010). Determination of minimal concentration of *Piscirickettsia salmonis* in water columns to establish a fallowing period in salmon farms. *Journal of Fish Diseases*, Vol.110, No.2, pp. 468-476, ISSN 1365-2761

Olson, T. K. & Criddle, K. R. (2008). Industrial Evolution: A Case Study of Chilean Salmon Aquaculture. *Aquaculture Economics & Management*, Vol.12, No.2, pp 89-106, ISSN 1365-7305

Ottem, K.; Nylund, A.; Karksbakk, E.; Friis-Møller, Kamaishi, T. (2009). Elevation of *Francisella philomiragia subsp. noatunensis* Mikalsen et al. (2007) to *Francisella noatunensis comb. nov.* [syn. *Francisella piscicida* Ottem et al. (2008) syn. nov] and characterization of *Francisella noatunensis susbsp. orientalis subsp. nov.*, two important fish pathogens. *Journal of Applied Microbiology*, Vol.106, No.4, pp. 1231-1243. ISSN 1365-2672

Perlman, H. and F. Juárez-Rubio (2010). Industrial Agglomerations: The Case of the Salmon Industry in Chile. *Aquaculture Economics & Management*, Vol.14, No.2, pp 164-184, ISSN 1365-7305

Reyes, X. & Bravo, S. (1983). Nota sobre una copepoidosis en salmones de cultivo. *Investigaciones marinas Valparaiso*, No. 11, pp. 55-57. ISSN 0716-1069

Rozas, M. & Ascencio, G. (2007). Evaluación de la situación epidemiológica de Caligiasis en Chile: hacia una estrategia de control efectiva. In: *Salmociencia*, Vol.2, No.2, pp. 43-59, ISSN 0718-5537

Rozas, M. (2011). Descripción patológica y epidemiológica de Amoebic Gill Disease (AGD) en salmón del Atlántico, *Salmo salar*, en Chile. Tesis de Magister en Ciencias Veterinarias. Universidad Austral de Chile, Valdivia.

SAG (2011). Productos biológicos inmunológicos con registro provisional uso en salmónidos, Servicio Agrícola y Ganadero. In: *http://www.sag.gob.cl*, 27.05.2011, Available from

    http://www.sag.gob.cl/common/asp/pagAtachadorVisualizador.asp?argCrypted
    Data=GP1TkTXdhRJAS2Wp3v88hIh98jlQCnaLAaTC9s9%2FJWY%3D&argModo=
    &argOrigen=BD&argFlagYaGrabados=&argArchivoId=37391

SalmonChile (2007a). Análisis estadístico y de Mercado, In: *www.salmonchile.cl*, 10.2010,
    Available from
    http://www.salmonchile.cl/frontend/seccion.asp?contid=117&secid=4&pag=1&s
    ubsecid=21

SalmonChile (2007b). La Contribución de la salmonicultura a la economía Chilena, In:
    www.salmonchile.cl, 10.2010, Available from
    http://www.salmonchile.cl/frontend/seccion.asp?contid=397&secid=33&secoldid
    =33&subsecid=117&pag=1

San Martín, B. (2010) Uno de los principales problemas es la escasez de arsenal terapéutico.
    In: *Revista Aqua*, 27.05.2011, Available from
    http://www.aqua.cl/entrevistas/entrevista.php?doc=294

San Martín, B.; Yatabe, T.; Gallardo, A. & Medina, P. (2010). Manual de Buenas Prácticas en
    el Uso de Antibióticos y Antiparasitarios en la Salmonicultura Chile Laboratorio de
    Farmacología Veterinaria, Universidad de Chile & Sernapesca. In
    http://www.sernapesca.cl, 27.05.2011, Available from
    http://www.sernapesca.cl/index.php?option=com_remository&Itemid=246&func
    =startdown&id=4097

Sepúlveda, M., F. Farías and E. Soto (2009). Salmon escapes in Chile. In: *Incidents, impacts,*
    mitigation and prevention, World Wildlife Fund (Ed.), 1-52, Valdivia, Chile.

Sernapesca (2008a) Unidad de Acuicultura Servicio Nacional de Pesca. In: *Informe sanitario de*
    la Acuicultura en Chile, 2007

Sernapesca (2008b). Balance de la situación sanitaria de la Anemia Infecciosa del Salmón en
    Chile de Julio del 2007 a Julio del 2008, Unidad de Acuicultura Servicio Nacional de
    Pesca. In *http://www.sernapesca.cl*, 27.05.2011, Available from
    http://www.sernapesca.cl/index.php?option=com_remository&Itemid=246&func
    =fileinfo&id=2659

Sernapesca (2009). Programa sanitario específico de vigilancia y control de Caligidosis.
    Resolución N° 2117 In *http://www.sernapesca.cl*, 27.05.2011, Available from
    http://www.sernapesca.cl/index.php?option=com_remository&Itemid=246&func
    =startdown&id=3702

Sernapesca (2010). Informe técnico: Resultados del diagnóstico general por jaula anual de
    Caligidosis en Chile (DGJA) de 2010, Unidad de Acuicultura Servicio Nacional de
    Pesca. In: *http://www.sernapesca.cl*, 27.05.2011, Available from
    http://www.sernapesca.cl/index.php?option=com_remository&Itemid=246&func
    =startdown&id=4481

Sernapesca (2011a). Solicitud de información pública Ley 20.285. Solicitud SIAC No.
    460131711, 10.03.2011, Valparaíso.

Sernapesca (2011b). Solicitud de información pública Ley 20.285. Solicitud SIAC
    N°460274611, 11.05.2011, Valparaíso

Sernapesca (2011c). Solicitud de información pública Ley 20.285. Solicitud SIAC No.
    460645210, 01.2011, Valparaíso

Sernapesca (2011d). Solicitud de información pública Ley 20.285. Solicitud SIAC No.
    460299811, 06.2011, Valparaíso

Silva, G (2010a). Producción de *smolt* en agua dulce: Los desafíos del confinamiento. *Aqua*
    Acuicultura + Pesca, Vol.22, No.145, pp 8-13

Silva, G. (2010b) Efectos del virus ISA: Una industria convaleciente. *Aqua Acuicultura +
Pesca*, Vol.22, No.139, pp. 28-33

Silva, G. (2010c) Producción de salmónidos 2009: El principio del ajuste. *Aqua Acuicultura +
Pesca*, Vol.22, No.139, pp. 8-16

Smith, P. A.; Larenas, J.; Contreras, J.; Cassigoli, J.; Venegas, C.; Rojas, M. E.; Guajardo, A.;
Pérez, S. & Díaz, S. (2006). Infectious haemolytic anemia causes jaundice outbreaks
in seawater–cultured coho salmon, *Oncorhynchus kisutch* (Walbaum), in Chile.
Journal of Fish Diseases, Vol.29, No.12, pp. 709-715, ISSN 1365-2761

Smith, P.; Contreras, J.; Larenas, J.; Aguillon, J.; Garces, L.; Perez, B. & Fryer J. (1997).
Immunization with bacterial antigens: piscirickettsiosis. *Developments in biological
standardization*, Vol.90, No.1, pp. 161-166, ISSN 0301-5149

Smith, P.; Larenas, J.; Vera, P.; Contreras, J.; Venegas C.; Rojas, M. & Guajardo, A. (2001).
Principales enfermedades de los peces salmonídeos cultivados en Chile.
Monografías de Medicina Veterinaria, Vol.21, No.2, pp 3-19, ISSN 0716-226X

Smith, P.A.; Larenas, J.; Contreras, J.; Cassigoli, J.; Venegas, C.; Rojas, M. E.; Guajardo, A.;
Troncoso, O. & Macías D. (2002). Infectious haemolytic anaemia of salmon: an
emerging disease occurring in seawater coho salmon (*Oncorhynchus kisutch*) in
Chile. *Proceedings of the 4th International Symposium of Aquatic Animal Health*, New
Orleans, September, 2002.

Subpesca (2010). Primer Informe Técnico: Propuesta de modificación D.S. 319 2001.
Reglamento de medidas de protección, control y erradicación de enfermedades de
alto riesgo para las especies hidrobiológicas, Subsecretaría de Pesca. In
http://www.subpesca.cl, 27.05.2011, Available from
http://www.subpesca.cl/controls/neochannels/neo_ch878/neochn878.aspx

Tapia, F. and S. Giglio (2010). Fjord Carrying Capacity Assessment Models Applicable to
Ecosystems in Southern Chile. In: *Technical Report*, World Wildlife Fund (Ed.), 1-22,
Valdivia, Chile.

TechnoPress (2010). Ranking of Chilean Salmon and Trout Exports, January-December 2009-
2010, In: *Directorio Aqua*, 03.05.2011, Available from
http://www.directorioaqua.com/contenido/dapel_esalmonidos.php

Troncoso, M.; Toledo, M. S.; Portell, P. & Figueroa G. (1994). Aislamiento de *Yersinia ruckeri* en
salmónidos de cultivo. *Avances en Ciencias Veterinarias*, Vol.9, No.2, ISSN 0716-260X

Valdebenito, S. & Avendaño-Herrera, R. (2009). Phenotypic, serological and genetic
characterization of *Flavobacterium psychrophilum* strains isolated from salmonids in
Chile. *Journal of Fish Diseases*, Vol.32, No.4, pp. 321–333, ISSN 1365-2761

Vilata, J.; Oliva, D. & Sepúlveda, M. (2010). The predation of farmed salmon by South
American sea lions (*Otaria flavescens*) in southern Chile. *ICES Journal of Marine
Science*, Vol.67, No.3, pp 475-482, ISSN 1095-9289

Permissions

The contributors of this book come from diverse backgrounds, making this book a truly international effort. This book will bring forth new frontiers with its revolutionizing research information and detailed analysis of the nascent developments around the world.

We would like to thank Barbara Sladonja, for lending her expertise to make the book truly unique. She has played a crucial role in the development of this book. Without her invaluable contribution this book wouldn't have been possible. She has made vital efforts to compile up to date information on the varied aspects of this subject to make this book a valuable addition to the collection of many professionals and students.

This book was conceptualized with the vision of imparting up-to-date information and advanced data in this field. To ensure the same, a matchless editorial board was set up. Every individual on the board went through rigorous rounds of assessment to prove their worth. After which they invested a large part of their time researching and compiling the most relevant data for our readers. Conferences and sessions were held from time to time between the editorial board and the contributing authors to present the data in the most comprehensible form. The editorial team has worked tirelessly to provide valuable and valid information to help people across the globe.

Every chapter published in this book has been scrutinized by our experts. Their significance has been extensively debated. The topics covered herein carry significant findings which will fuel the growth of the discipline. They may even be implemented as practical applications or may be referred to as a beginning point for another development. Chapters in this book were first published by InTech; hereby published with permission under the Creative Commons Attribution License or equivalent.

The editorial board has been involved in producing this book since its inception. They have spent rigorous hours researching and exploring the diverse topics which have resulted in the successful publishing of this book. They have passed on their knowledge of decades through this book. To expedite this challenging task, the publisher supported the team at every step. A small team of assistant editors was also appointed to further simplify the editing procedure and attain best results for the readers.

Our editorial team has been hand-picked from every corner of the world. Their multi-ethnicity adds dynamic inputs to the discussions which result in innovative outcomes. These outcomes are then further discussed with the researchers and contributors who give their valuable feedback and opinion regarding the same. The feedback is then collaborated with the researches and they are edited in a comprehensive manner to aid the understanding of the subject.

Apart from the editorial board, the designing team has also invested a significant amount of their time in understanding the subject and creating the most relevant covers. They scrutinized every image to scout for the most suitable representation of the subject and create an appropriate cover for the book.

The publishing team has been involved in this book since its early stages. They were actively engaged in every process, be it collecting the data, connecting with the contributors or procuring relevant information. The team has been an ardent support to the editorial, designing and production team. Their endless efforts to recruit the best for this project, has resulted in the accomplishment of this book. They are a veteran in the field of academics and their pool of knowledge is as vast as their experience in printing. Their expertise and guidance has proved useful at every step. Their uncompromising quality standards have made this book an exceptional effort. Their encouragement from time to time has been an inspiration for everyone.

The publisher and the editorial board hope that this book will prove to be a valuable piece of knowledge for researchers, students, practitioners and scholars across the globe.

List of Contributors

César Alejandro Berlanga-Robles and Arturo Ruiz-Luna
Centro de Investigación en Alimentación y Desarrollo A. C., Unidad Regional Mazatlán, Mexico

Rafael Hernández-Guzmán
Posgrado en Ciencias del Mar y Limnología, UNAM, Mexico

Magdalena Lagunas-Vazques, Giovanni Malagrino and Alfredo Ortega-Rubio
Centro de Investigaciones Biológicas del Noroeste, La Paz, Baja California Sur., Mexico

Maricar S. Samson
Br. Alfred Shields FSC Marine Station, De La Salle University, Manila, Philippines

Maricar S. Samson
School of Environmental Science and Management, University of the Philippines Los Baños, Philippines

Rene N. Rollon
Institute of Environmental Science and Meteorology, University of the Philippines Diliman, Philippines

Barbara Sladonja
Institute of Agriculture and Tourism Poreč, Poreč, Croatia

Nicola Bettoso, Francesco Tamberlich and Alessandro Acquavita
ARPA FVG-Osservatorio Alto Adriatico, Trieste, Italy

Aurelio Zentilin
Almar Soc. Coop. Agricola a.r.l., I-33050 Marano Lagunare (UD), Italy

Yan Liu, Guohai Dong, Yunpeng Zhao and Yucheng Li
Dalian University of Technology, China

Changtao Guan
Yellow Sea Fisheries Research Institute, China

Changtao Guan
Key Laboratory of Fishery Equipment and Engineering, China

Oscar Alatorre-Jácome, Fernando García-Trejo, Enrique Rico-García and Genaro M. Soto-Zarazúa
C.A. de Ingeniería de Biosistemas, Campus Amazcala, Facultad de Ingeniería, Universidad Autónoma de Querétaro, Amazcala, Mpio, El Marqués,C.P.76260, Querétaro, Qro, México

Salud Deudero, Ariadna Tor, Carme Alomar and Piluca Sarriera
Centro Oceanográfico de Baleares, Instituto Español de Oceanografia, Palma, Spain

José Maria Valencia
Laboratorio de Investigaciones Marinas y Acuicultura, Andratx, Spain

Andreu Blanco
Instituto de Investigaciones Marinas de Vigo, CSIC, Vigo, Spain

Simonel Sandu and Eric Hallerman
Virginia Polytechnic Institute and State University, USA

Brian Brazil
U.S. Department of Agriculture – Agricultural Research Service, USA

Brian Brazil
Geosyntec Consultants, N.W., Kennesaw, GA, USA

Joseph Sneddon and Joel C. Richert
Department of Chemistry, McNeese State University, Lake Charles, Louisiana, USA

Carlos Rosas, Cristina Pascual and Maite Mascaró
Unidad multidisciplinaria de Docencia e Investigación, Fac. de Ciencias, UNAM, Puerto de abrigo s/n, Sisal, Yucatán, México

Paulina Gebauer
Centro I-Mar. Universidad de los Lagos, Puerto Montt, Chile, Chile

Ana Farias, Kurt Paschke and Iker Uriarte
Instituto de Acuicultura, Universidad Austral, de Chile, Sede Puerto Montt, Chile

Ana Farias and Iker Uriarte
CIEN Austral, Chile

Pablo Ibieta, Valentina Tapia, Claudia Venegas, Mary Hausdorf and Harald Takle
AVS Chile SA, Puerto Varas, Chile

Harald Takle
Nofima, Ås, Norway